UNITED NATIONS CONFERENCE ON TRADE AND DEVELOPMENT

REVIEW
OF MARITIME TRANSPORT
2015

UNITED NATIONS
New York and Geneva, 2015

NOTE

The *Review of Maritime Transport* is a recurrent publication prepared by the UNCTAD secretariat since 1968 with the aim of fostering the transparency of maritime markets and analysing relevant developments. Any factual or editorial corrections that may prove necessary, based on comments made by Governments, will be reflected in a corrigendum to be issued subsequently.

* * *

Symbols of United Nations documents are composed of capital letters combined with figures. Use of such a symbol indicates a reference to a United Nations document.

* * *

The designations employed and the presentation of the material in this publication do not imply the expression of any opinion whatsoever on the part of the United Nations concerning the legal status of any country, territory, city or area, or of its authorities, or concerning the delimitation of its frontiers or boundaries.

* * *

Material in this publication may be freely quoted or reprinted, but acknowledgement is requested, with reference to the document symbol (UNCTAD/RMT/2015). A copy of the publication containing the quotation or reprint should be sent to the UNCTAD secretariat at the following address: Palais des Nations, CH-1211 Geneva 10, Switzerland.

UNCTAD/RMT/2015

UNITED NATIONS PUBLICATION

Sales no. E. 15.II.D.6

ISBN 978-92-1-112892-5

eISBN: 978-92-1-057410-5

ISSN 0566-7682

ACKNOWLEDGEMENTS

The *Review of Maritime Transport 2015* has been prepared by UNCTAD. The preparation was coordinated by Jan Hoffmann with administrative support and formatting by Wendy Juan, under the overall guidance of Anne Miroux. Contributors were Regina Asariotis, Hassiba Benamara, Jan Hoffmann, Anila Premti, Ricardo Sanchez, Vincent Valentine, Gordon Wilmsmeier and Frida Youssef.

The publication was edited by Deniz Barki and John Rogers. The cover was designed by Sophie Combette. The desktop publishing was carried out by Nathalie Loriot.

The considered comments and valuable input provided by the following reviewers are gratefully acknowledged: Celine Bacrot, James Coldwell, Trevor Crowe, Mahin Faghfouri, Peter Faust, Marco Fugazza, Ki-Soon Hwang, Nicolas Maystre, Shin Ohinata, Tansuğ Ok, Richard Oloruntoba, Christopher Pålsson, Sarinka Parry-Jones, Dong-Wook Song, Patricia Sourdin and André Stochniol. Thanks are also due to Vladislav Shuvalov for reviewing the publication in full.

TABLE OF CONTENTS

Note ii
Acknowledgements iii
List of tables, figures and boxes v
Abbreviations vii
Explanatory notes viii
Vessel groupings used in the *Review of Maritime Transport* ix
Executive summary x

1. DEVELOPMENTS IN INTERNATIONAL SEABORNE TRADE 1

    A. World economic situation and prospects 2
    B. World seaborne trade 5
    C. Sustainable and resilient maritime transport systems 21

2. STRUCTURE, OWNERSHIP AND REGISTRATION OF THE WORLD FLEET 29

    A. Structure of the world fleet 30
    B. Ownership and operation of the world fleet 35
    C. Container ship deployment and liner shipping connectivity 39
    D. Registration of ships 41
    E. Shipbuilding, demolition and new orders 43

3. FREIGHT RATES AND MARITIME TRANSPORT COSTS 47

    A. Determinants of maritime transport costs 48
    B. International transport costs 54
    C. Recent developments in freight rates 56

4. PORTS 65

    A. Ports and port-related developments 66
    B. International terminal operators 70
    C. Sustainability challenges facing ports 73

5. LEGAL ISSUES AND REGULATORY DEVELOPMENTS 79

    A. Important developments in transport law 80
    B. Regulatory developments relating to the reduction of greenhouse gas emissions from international shipping and other environmental issues 83
    C. Other legal and regulatory developments affecting transportation 89
    D. Status of conventions 96
    E. Trade facilitation and sustainable development 97

# LIST OF TABLES, FIGURES AND BOXES

Tables

1.1.	World economic growth, 2012–2015 (annual percentage change)	2
1.2.	Growth in the volume of merchandise, 2012–2014 (annual percentage change)	4
1.3.	Developments in international seaborne trade, selected years (millions of tons loaded)	6
1.4 (a).	World seaborne trade 2006–2014, by type of cargo, country group and region (millions of tons)	8
1.4 (b).	World seaborne trade 2006–2014, by type of cargo, country group and region (percentage share)	10
1.5.	Major producers and consumers of oil and natural gas, 2014 (world market share in percentage)	15
1.6.	Some major dry bulks and steel: Main producers, users, exporters and importers, 2014 (world market shares in percentages)	17
1.7.	Estimated containerized cargo flows on major East–West container trade routes, 2009–2014 (million TEUs and percentage annual change)	21
2.1.	World fleet by principal vessel types, 2014–2015 (beginning-of-year figures, thousands of dwt; percentage share in italics)	31
2.2.	Age distribution of the world merchant fleet, by vessel type, as of 1 January 2015 (percentage of total ships and dwt)	33
2.3.	Ownership of the world fleet, as of 1 January 2015 (dwt)	36
2.4.	The 50 leading liner companies, 1 May 2015 (Number of ships and total shipboard capacity deployed, ranked by TEU)	37
2.5.	Container ship deployment on selected routes, 1 May 2015	40
2.6.	The 35 flags of registration with the largest registered fleets, as of 1 January 2015 (dwt)	42
2.7.	Distribution of dwt capacity of vessel types, by country group of registration, January 2015 (beginning-of-year figures, per cent of dwt, annual growth in percentage points in italics)	43
2.8.	Deliveries of newbuildings, major vessel types and countries where built (2014, thousands of GT)	44
2.9.	Tonnage reported sold for demolition, major vessel types and countries where demolished (2014, thousands of GT)	44
3.1.	Container freight markets and rates	58
3.2.	Baltic Exchange tanker indices	59
3.3.	Tanker market summary – clean and dirty spot rates, 2010–2014 (Worldscale)	60
4.1.	Container port throughput for 80 developing countries/territories and economies in transition, 2012–2014 (TEUs)	67
4.2.	Top 20 container terminals and their throughput, 2012–2014 (TEUs and percentage change)	69
4.3.	Top global terminals' berth productivity, 2014 (container moves per ship, per hour on all vessel sizes)	70
4.4.	World's leading ports by productivity, 2014 (container moves per ship, per hour on all vessel sizes and percentage increase)	71
5.1.	Contracting States Parties to selected international conventions on maritime transport as at 30 June 2015	97
5.2.	Examples of articles of the TFA that may benefit from and help to achieve SDGs	98

Figures

1.1.	The OECD Industrial Production Index and indices for world GDP, merchandise trade and seaborne shipments (1975–2014) (base year 1990 = 100)	5
1.2.	International seaborne trade, selected years (millions of tons loaded)	6
1.3.	Structure of international seaborne trade, 2014	7
1.4 (a).	World seaborne trade, by country group, 2014 (percentage share in world tonnage)	12
1.4 (b).	Participation of developing countries in world seaborne trade, selected years (percentage share in world tonnage)	12
1.4 (c).	World seaborne trade, by region, 2014 (percentage share in world tonnage)	13
1.5.	World seaborne trade in cargo ton–miles by cargo type, 2000–2015 (billions of ton–miles)	15
1.6 (a).	Global containerized trade, 1996–2015 (million TEUs and percentage annual change)	19
1.6 (b).	Estimated containerized cargo flows on major East–West container trade routes (million TEUs), 1995–2014	20
1.6 (c).	Distribution of global containerized trade by route, 2014 (percentage share of global trade in TEUs)	20
2.1.	Annual growth of the world fleet, 2000–2014 (per cent of dwt)	30
2.2.	World fleet by principal vessel types, 1980–2015 (beginning-of-year figures, percentage share of dwt)	31
2.3.	Contract year for tonnage (dwt) delivered in 2014	32
2.4.	Vessel types of the world fleet, by year of building (dwt as of 1 January 2015)	32
2.5.	Share of newbuildings (number of ships) with ballast water treatment systems, by main vessel type, 2007–2014	35
2.6.	Presence of liner shipping companies: Average number of companies per country and average container-carrying capacity deployed (TEUs) per company per country (2004–2015)	41
2.7.	World tonnage on order, 2000–2015 (thousands of dwt)	45
3.1.	Determinants of maritime transport costs	48
3.2.	Statistical correlation between articles of the WTO TFA and indicators for trade efficiency	49
3.3.	The "no-relationship" between distance and maritime transport costs	50
3.4.	The relationship between transport costs and LSBCI, 2012 and 2013	51
3.5.	Transport costs and economies of scale	52
3.6.	International transport costs: Freight costs as a percentage of value of imports, ten-year averages within country groups, 1985–2014	55
3.7.	Growth of demand and supply in container shipping, 2000–2015 (annual growth rates)	57
3.8.	Baltic Exchange Dry Index, 2012–2015 (index base year 1985 = 1,000 points)	61
3.9.	Daily earnings of bulk carrier vessels, 2008–2015 ($ per day)	62
5.1.	The Human Development Index (HDI) and the number of trade facilitation measures notified as category A	99

Boxes

1.1.	Examples of voluntary self-regulation in shipping	22
5.1.	The current status of the ISO 28000 series of standards	92

ABBREVIATIONS

AEO	authorized economic operator
BWM Convention	International Convention for the Control and Management of Ships' Ballast Water and Sediments
CBP	Customs and Border Protection (United States of America)
CH_4	methane
CO	carbon monoxide
CO_2	carbon dioxide
COP21	twenty-first session of the Conference of the Parties to the United Nations Framework Convention on Climate Change
CSAV	Compañía Sudamericana de Vapores
CSI	Container Security Initiative
C–TPAT	Customs–Trade Partnership against Terrorism (United States of America)
dwt	dead-weight ton
ECA	emission control area
ECLAC	Economic Commission for Latin America and the Caribbean
EEDI	Energy Efficiency Design Index
FEU	40-foot equivalent unit
FPSO	floating production, storage and offloading unit
GDP	gross domestic product
GHG	greenhouse gas
GT	gross tonnage
HDI	Human Development Index
HNS	hazardous noxious substances
HNS Convention	International Convention on Liability and Compensation for Damage in Connection with the Carriage of Hazardous and Noxious Substances by Sea
IAPH	International Association of Ports and Harbors
ILO	International Labour Organization
IMO	International Maritime Organization
ISO	International Organization for Standardization
ISPS Code	International Ship and Port Facilities Security Code
JOC	*Journal of Commerce*
$kgCO_2e/modTEU$	kilograms CO_2 emitted per modified 20-foot equivalent unit
LDC	least developed country
LNG	liquefied natural gas
LPG	liquefied petroleum gas
LPI	Logistics Performance Index (World Bank)
LSBCI	Liner Shipping Bilateral Connectivity Index (UNCTAD)
LSCI	Liner Shipping Connectivity Index (UNCTAD)
MARPOL	International Convention for the Prevention of Pollution from Ships
MEPC	Marine Environment Protection Committee (IMO)
MLC	Maritime Labour Convention
MRA	mutual recognition agreement
MSC	Maritime Safety Committee (IMO)
N_2O	nitrous oxide
NO_x	nitrogen oxides
OECD	Organization for Economic Cooperation and Development
PM	particulate matter
ppm	parts per million

SAFE	Framework of Standards to Secure and Facilitate Global Trade
SDG	sustainable development goal
SEEMP	Ship Energy Efficiency Management Plan
SID	seafarers' identity document
SIDS	small island developing State(s)
SOLAS	International Convention for the Safety of Life at Sea
SO_2	sulphur dioxide
SO_x	sulphur oxides
STCW	International Convention on Standards of Training, Certification and Watchkeeping for Seafarers
TEU	20-foot equivalent unit
TFA	Trade Facilitation Agreement (World Trade Organization)
UNCLOS	United Nations Convention on the Law of the Sea
UNCTAD	United Nations Conference on Trade and Development
UNDP	United Nations Development Programme
UNFCCC	United Nations Framework Convention on Climate Change
WCO	World Customs Organization
WTO	World Trade Organization

EXPLANATORY NOTES

- The *Review of Maritime Transport 2015* covers data and events from January 2014 until June 2015. Where possible, every effort has been made to reflect more recent developments;

- All references to dollars ($) are to United States of America dollars, unless otherwise stated;

- Unless otherwise stated, "ton" means metric ton (1,000 kg) and "mile" means nautical mile;

- Because of rounding, details and percentages presented in tables do not necessarily add up to the totals;

- n.a.: not available;

- A hyphen (-) signifies that the amount is nil;

- In the tables and the text, the terms "countries" and "economies" refer to countries, territories or areas;

- Since 2014, the *Review of Maritime Transport* does not include printed statistical annexes. Instead, UNCTAD has expanded the coverage of statistical data on compact disc and online via the following links:

    Seaborne trade: http://stats.unctad.org/seabornetrade

    Merchant fleet by flag of registration: http://stats.unctad.org/fleet

    Merchant fleet by country of ownership: http://stats.unctad.org/fleetownership

    Merchant fleet by country of ownership and flag of registration: http://stats.unctad.org/shipregistration

    Ship building by country in which built: http://stats.unctad.org/shipbuilding

    Ship scrapping by country of demolition: http://stats.unctad.org/shipscrapping

    Liner Shipping Connectivity Index (LSCI): http://stats.unctad.org/lsci

    Liner Shipping Bilateral Connectivity Index (LSBCI): http://stats.unctad.org/lsbci

    Containerized port traffic: http://stats.unctad.org/teu

Vessel groupings used in the *Review of Maritime Transport*

Group	Constituent ship types
Oil tankers	Oil tankers
Bulk carriers	Bulk carriers, combination carriers
General-cargo ships	Multi-purpose and project vessels, roll-on roll-off cargo, general cargo
Container ships	Fully cellular container ships
Other ships	Liquefied petroleum gas carriers, liquefied natural gas carriers, parcel (chemical) tankers, specialized tankers, reefers, offshore supply ships, tugs, dredgers, cruise ships, ferries, other non-cargo ships
Total all ships	Includes all the above-mentioned vessel types

Approximate vessel size groups referred to in the *Review of Maritime Transport*, according to generally used shipping terminology

Crude oil tankers

Very large crude carrier	200,000 dead-weight tons (dwt) plus
Suezmax crude tanker	120,000–200,000 dwt
Aframax crude tanker	80,000–119,999 dwt
Panamax crude tanker	60,000–79,999 dwt

Dry bulk and ore carriers

Capesize bulk carrier	100,000 dwt plus
Panamax bulk carrier	60,000–99,999 dwt
Handymax bulk carrier	40,000–59,999 dwt
Handysize bulk carrier	10,000–39,999 dwt

Container ships

Post-Panamax container ship	beam of > 32.3 metres
Panamax container ship	beam of < 32.3 metres

Source: Clarksons Research.

Note: Unless otherwise specified, the ships covered in the *Review of Maritime Transport* include all propelled seagoing merchant vessels of 100 gross tonnage (GT) and above, excluding inland waterway vessels, fishing vessels, military vessels, yachts and offshore fixed and mobile platforms and barges (with the exception of floating production storage and offloading units (FPSOs) and drillships).

EXECUTIVE SUMMARY

The year 2015 is a milestone for sustainable development. The international community has a unique opportunity to strengthen its commitment to sustainable development and consider how best to mainstream sustainability principles across all economic activities and sectors, including maritime transport. In this context, relevant chapters of the present edition of the *Review of Maritime Transport* highlight some issues that are at the interface of maritime transport and sustainability and underscore the role of maritime transport in helping implement a workable international sustainable development agenda.

Seaborne trade

The world economy embarked on a slow-moving recovery led by uneven growth in developed economies and a slowdown in developing countries and economies in transition. In 2014, the world gross domestic product (GDP) increased marginally by 2.5 per cent, up from 2.4 per cent in 2013. Meanwhile, world merchandise trade increased by 2.3 per cent; this is down from 2.6 per cent in 2013 and below the pre-crisis levels.

Accordingly, preliminary UNCTAD estimates indicate that global seaborne shipments have increased by 3.4 per cent in 2014, that is at the same rate as in 2013. Additions to volumes exceeded 300 million tons taking the total to 9.84 billion tons. This performance unfolded in the context of a number of developments, including (a) a slowdown in large emerging developing economies; (b) lower oil price levels and new refinery capacity developments; and (c) a slow-moving and uneven recovery in the advanced economies.

On balance, growth in world GDP, merchandise trade and seaborne shipments is expected to continue at a moderate pace in 2015. The outlook remains uncertain and subject to many downside risks, including continued moderate growth in global demand and merchandise trade, the fragile recovery in Europe, diverging outlooks for net oil consumers and producers, geopolitical tensions, and a potential faster slowdown in developing economies, in particular the large emerging economies, as well as uncertainty about the pace and the implications of the slowdown in China.

The fleet

The world fleet grew by 3.5 per cent during the 12 months to 1 January 2015, the lowest annual growth rate in over a decade. In total, at the beginning of the year, the world's commercial fleet consisted of 89,464 vessels, with a total tonnage of 1.75 billion dwt. For the first time since the peak of the shipbuilding cycle, the average age of the world fleet increased slightly during 2014. Given the delivery of fewer newbuildings, combined with reduced scrapping activity, newer tonnage no longer compensated for the natural aging of the fleet.

Greece continues to be the largest ship-owning country, followed by Japan, China, Germany and Singapore. Together, the top five ship-owning countries control more than half of the world tonnage. Five of the top 10 ship-owning countries are from Asia, four are European and one is from the Americas.

The *Review of Maritime Transport* further illustrates the process of concentration in liner shipping. While the container-carrying capacity per provider per country tripled between 2004 and 2015, the average number of companies that provide services from/to each country's ports decreased by 29 per cent. Both trends illustrate two sides of the same coin: as ships get bigger and companies aim at achieving economies of scale, there remain fewer companies in individual markets.

New regulations require the shipping industry to invest in environmental technologies, covering issues such as emissions, waste, and ballast water treatment. Some of the investments are not only beneficial for the environment, but may also lead to longer-term cost savings, for example due to increased fuel efficiency.

Economic and regulatory incentives will continue to encourage individual owners to invest in modernizing their fleets. Unless older tonnage is demolished, this would lead to further global overcapacity, continuing the downward pressure on freight and charter rates. The interplay between more stringent environmental regulations and low freight and charter rates should encourage the further demolition of older vessels; this will not only help reduce the oversupply in the market, but also contribute to lowering the global environmental impact of shipping.

Freight costs

Developing countries, especially in Africa and Oceania, pay 40 to 70 per cent more on average for the international transport of their imports than developed countries. The main reasons for this situation are to be found in these regions' trade imbalances, pending port and trade facilitation reforms, as well as lower trade volumes and shipping connectivity. There is potential for policymakers to partly remedy the situation through investments and reforms, especially in the regions' seaports, transit systems and customs administrations.

Container freight rates remained volatile throughout 2014 although with different trends on individual trade lanes. Market fundamentals have not changed significantly despite the expansion in global demand for container shipping. This was mainly due to pressure from the constant supply of vessels that the market rates continued to face, with the introduction of very large units on mainlane trades and the cascading effect on non-mainlanes trades. The tanker market, which encompasses the transportation of crude oil, refined petroleum products and chemicals, witnessed an equally volatile freight rate environment in 2014 and early 2015. The dry bulk market freight rates faced another challenging year influenced by the surplus capacity that still exists and the uncertainties in demand projections. Bulk carrier earnings fell 5 per cent from 2013 to reach an average of $9,881 per day in 2014. The low level of earnings exerted financial pressure on owners and led to several companies filing for bankruptcy.

Ports

Developing economies' share of world container port throughput increased marginally to approximately 71.9 per cent. This continues the trend of a gradual rise in developing countries' share of world container throughput. The increased share of world container throughput for developing countries reflects an increase in South–South trade.

The performance of ports and terminals is important because it affects a country's trade competitiveness. There are many determinants to port/terminal performance – labour relations, number and type of cargo handling equipment, quality of backhaul area, port access channel, land-side access and customs efficiency, as well as potential concessions to international terminal operators. The world's largest terminal operator handled 65.4 million 20-foot equivalent units (TEUs) in 2014, an increase of 5.5 per cent over the previous year. Of this figure, 33.6 million TEUs related to its operations at the port of Singapore and 31.9 million TEUs from its international portfolio. Hutchison Port Holdings trust is the second largest international terminal operator by market share. With operations in China and Hong Kong, China, it is not as geographically diverse as some other international terminal operators. APM Terminals has a geographical presence in 39 countries. DP World is the most geographically diverse of the global terminal operators, with a network of more than 65 terminals spanning six continents.

The economic, environmental and social challenges facing ports include growing and concentrated traffic volumes brought about by ever-increasing ship size; the cost of adaptation of port and port hinterland infrastructure measures; a changing marketplace as a result of increased alliances between shipping lines; national budget constraints limiting the possibilities of public funding for transport infrastructure; volatility in energy prices, the new energy landscape and the transition to alternative fuels; the entry into force of stricter sulphur limits (in, for example, International Maritime Organization (IMO) emission control area (ECA) countries); increasing societal and environmental pressure; and potential changes in shipping routes from new or enlarged international passage ways.

Legal and regulatory framework

In 2014, important regulatory developments in the field of transport and trade facilitation included the adoption of the International Code for Ships Operating in Polar Waters (Polar Code), expected to enter into force on 1 January 2017, as well as a range of regulatory developments relating to maritime and supply chain security and environmental issues.

To further strengthen the legal framework relating to ship-source air pollution and the reduction of greenhouse gas (GHG) emissions from international shipping, several regulatory measures were adopted at IMO, and the third IMO GHG Study 2014 was finalized. Also, guidelines for the development of the Inventory of Hazardous Materials required under the 2010 International Convention on Liability and Compensation for Damage in Connection with the Carriage of Hazardous and Noxious Substances by

Sea (HNS Convention) – which, however, is not yet in force – were adopted, and further progress was made with respect to technical matters related to ballast water management, ship recycling, and measures helping to prevent and combat pollution of the sea from oil and other harmful substances.

Continued enhancements were made to regulatory measures in the field of maritime and supply chain security and their implementation, including the issuance of a new version of the World Customs Organization (WCO) Framework of Standards to Secure and Facilitate Global Trade (SAFE Framework) in June 2015, which includes a new pillar 3: "Customs-to-other government and inter-government agencies". As regards suppression of maritime piracy and armed robbery, positive developments were noted in the waters off the coast of Somalia and the wider western Indian Ocean. However, concern remains about the seafarers still being held hostage. A downward trend of attacks in the Gulf of Guinea was also observed, indicating that international, regional and national efforts are beginning to take effect.

DEVELOPMENTS IN INTERNATIONAL SEABORNE TRADE

The world economy embarked on a slow-moving recovery led by uneven growth in developed economies and a slowdown in developing countries and economies in transition. In 2014, the world gross domestic product (GDP) increased marginally by 2.5 per cent, up from 2.4 per cent in 2013. Meanwhile, world merchandise trade increased by 2.3 per cent; this is down from 2.6 per cent in 2013 and below the pre-crisis levels.

Accordingly, preliminary UNCTAD estimates indicate that global seaborne shipments have increased by 3.4 per cent in 2014, that is at the same rate as in 2013. Additions to volumes exceeded 300 million tons taking the total to 9.84 billion tons. This performance unfolded in the context of a number of developments, including (a) a slowdown in large emerging developing economies; (b) lower oil price levels and new refinery capacity developments; and (c) a slow-moving and uneven recovery in the advanced economies.

On balance, growth in world GDP, merchandise trade and seaborne shipments is expected to continue at a moderate pace in 2015. The outlook remains uncertain and subject to many downside risks, including continued moderate growth in global demand and merchandise trade, the fragile recovery in Europe, diverging outlooks for net oil consumers and producers, geopolitical tensions, and a potential faster slowdown in developing economies, in particular the large emerging economies, as well as uncertainty about the pace and the implications of the slowdown in China.

A. WORLD ECONOMIC SITUATION AND PROSPECTS

1. World economic growth

Global GDP increased by 2.5 per cent in 2014, up from 2.4 per cent in 2013 (see table 1.1). Although positive, this growth remains below the pre-crisis levels with almost all economies having shifted to a lower growth path. Growth in the advanced economies accelerated to 1.6 per cent, while GDP in both the developing economies and the economies in transition expanded at the slower rates of 4.5 per cent and 0.9 per cent, respectively. The emerging recovery in the advanced economies was uneven, led by accelerated growth in the United States (2.4 per cent) and the United Kingdom of Great Britain and Northern Ireland (3.0 per cent) and a fragile recovery in the European Union (1.3 per cent). Meanwhile, GDP growth in Japan came to a standstill due, among other factors, to the 2014 consumption tax increase and the fading away of the effect of the fiscal and monetary stimulus introduced in 2013.

Gross domestic product growth in the transition economies was constrained by weak exports and external financing constraints as well as the uncertainty caused by the geopolitical conflicts in the region. Although developing countries remained the engine of growth, contributing three quarters of global expansion in 2014 (International Monetary Fund, 2015), slower GDP growth reflects, in particular, weaker expansion in developing America and a slowdown in China. Elsewhere, the economies of the least developed countries (LDCs) continued to expand at a rapid rate (5.3 per cent).

China continued to grow at the relatively robust rate of 7.4 per cent. However, this rate is much below the average growth of 10.0 per cent achieved years earlier and reflects, to a large extent, the slowdown in the industrial production. Growth in industrial production averaged 8.0 per cent in 2014, down from 14.0 per cent in 2011 and 10 per cent in 2012 and 2013 (*Dry Bulk Trade Outlook*, 2015a). Meanwhile, GDP in India expanded by 7.1 per cent and is expected to grow at a faster rate in 2015. The slowdown in China entails some important implications for seaborne trade,

Table 1.1. World economic growth, 2012–2015 (annual percentage change)

Region/country	2012	2013	2014	2015[a]
WORLD	2.2	2.4	2.5	2.5
Developed economies	1.1	1.3	1.6	1.9
of which:				
European Union 28	-0.5	0.1	1.3	1.7
of which:				
France	0.2	0.7	0.2	1.2
Germany	0.4	0.1	1.6	1.5
Italy	-2.8	-1.7	-0.4	0.7
United Kingdom	0.7	1.7	3.0	2.3
Japan	1.7	1.6	-0.1	0.9
United States	2.3	2.2	2.4	2.3
Developing economies	4.7	4.8	4.5	4.1
of which:				
Africa	5.1	3.8	3.4	3.2
South Africa	2.2	2.2	1.5	1.9
Asia	5.1	5.6	5.5	5.2
China	7.7	7.7	7.4	6.9
India	4.4	6.4	7.1	7.5
Western Asia	4.0	4.1	3.3	2.5
Developing America	3.2	2.8	1.4	0.8
Brazil	1.8	2.7	0.1	-1.5
Least developed countries	4.3	5.3	5.3	3.5
Transition economies	3.3	2.0	0.9	-2.6
of which:				
Russian Federation	3.4	1.3	0.6	-3.5

Source: UNCTAD. *Trade and Development Report 2015*. Table 1.1.
[a] Forecast.
Note: Calculations for country aggregates are based on GDP at constant 2005 dollars.

shipping investors, service providers and users in view of the country's major role in supporting growth in Asia as well as in other developing regions. On the import side, dry bulk shipping and crude oil tankers have benefited the most from China's robust demand while, on the export side, container shipping, especially on the intra-Asian routes and westbound to Europe and North America, was the main beneficiary. The impact of a further slowdown in China will extend beyond the Chinese and Asian borders.

Looking forward, global economic growth is projected to moderate in 2015 supported mainly by growth in the advanced economies and relatively strong growth in Asia. Growth in developing countries as a group is expected to decelerate due to factors such as the low oil price levels and their impact on oil exporting countries, persistent political uncertainties, concerns about developments involving the European Union and Greece, and a continued rebalancing of China's economy.

The precise impact of lower oil prices will depend largely on their duration. The broad effects of a drop are generally positive as it stimulates global demand. However, this also implies an income shift from oil producers to consumers. Lower oil price levels will support the purchasing power of consumers in importing countries. For example, a sustained $30 decline in oil prices is expected to result in over $200 billion per year of savings for consumers in the United States through lower prices for gasoline, diesel, jet fuel and home heating oil (*Politico Magazine*, 2014). Conversely, demand from oil exporting countries will be constrained, including as a result of fiscal adjustments (for example, cuts of subsidies), unfavourable terms of trade and loss of revenue. It is estimated that each one-dollar fall in oil prices will result in a $2 billion loss in revenue for the Russian Federation (Johnson, 2015). Meanwhile, the oil and gas export earnings of the Gulf Cooperation Council countries are expected to decline by around $300 billion (International Monetary Fund, 2015). Other potential impacts of persistent lower oil prices relate to the delays, postponements or cancellations of oil and gas investment projects that may only have been feasible in a higher energy price-setting. Reduced energy sector investments will, in the medium or long term, likely dampen production as well as growth in oil and gas trades.

In sum, the world economy has embarked on a slow moving global recovery. On balance, GDP growth is expected to continue to moderate in 2015 with the outlook remaining subject to many downside risks, including a global demand and merchandise trade that undershoot expectations, the different economic outlooks for net oil consumers and producers, political shocks and geopolitical tensions, a potential faster slowdown in large developing economies, as well as uncertainty about the pace of the slowdown in China and related implications for the world economy, trade and seaborne shipments.

2. World merchandise trade

In 2014, the volume of global merchandise trade (that is, trade in value terms but adjusted to account for inflation and exchange rate movements) increased at the slower rate of 2.3 per cent, down from 2.6 per cent in 2013. Reflecting an uneven recovery in the advanced economies, this performance remained below the pre-crisis trends, with slower growth in developing economies and economies in transition constraining growth in overall merchandise trade volumes (see table 1.2).

Despite the deceleration recorded in 2014, developing countries continue to fuel global merchandise trade flows. UNCTAD data indicate that although developed economies continue to contribute the largest shares to world exports and imports (51.1 per cent and 54.9 per cent, respectively, in 2014), their contribution has been declining over the years. Meanwhile, the contribution of developing countries and economies in transition to world merchandise trade has been on the rise.

The share of developing countries in world exports in 2014 was estimated at 45.0 per cent (32.0 per cent in 2000), while their share of world imports amounted to 42.2 per cent (28.9 per cent in 2000). This reflects the shift in economic influence observed over recent years whereby developing countries are gaining greater market share in world merchandise trade both in terms of growth and levels.

The uneven performances among and within country groupings impacted the performance of containerized trade in 2014. Breaking away from patterns observed since 2009, volumes on the Asia–Europe and trans-Pacific container trade lanes (peak leg) reversed trends and recorded robust growth during the year.

Projected growth remains vulnerable, however, given continued uncertainties arising in connection with weaker growth in emerging economies, particularly a potential sharp slowdown in China, as well as concerns about the fragile recovery in the European Union and the situation in Greece. The slowdown in China will impact on the global recovery in trade volumes and affect the prospects of other countries, especially developing countries that have over recent years deepened their economic and trade relations with China through greater integration into regional and global value chains and by emerging as key sources of supply in terms of raw commodities.

A rebalancing of China's economy can significantly reshape the maritime transport landscape and alter shipping and seaborne trade patterns. The super cycle experienced by shipping over the past years was driven by globalization and rapid growth in the division of international labour and fragmentation in international production processes. Within the globalized context, the resource-intensive growth phase of China and its greater integration into the global production and value chains have been a key driver. As China has generated much of the growth in world seaborne trade since 2009, the challenge for shipping is to ensure that the trade dynamism generated by China's expansion continues and is replicated elsewhere.

In addition to the performance of global GDP and trade, other factors may also be at play and currently shaping the slow global economic and trade recovery. The long-term trade to GDP ratio of two to one appears to be unwinding. Over the past few years, world GDP has been growing at about the same rate as trade. This may be the result of limited growth in the fragmentation of global production processes, a maturation of value chains (in China and the United States), a change in the composition of global demand with a slow recovery in investment goods that are more trade intensive than government and consumer spending, costlier or limited trade finance, and potentially a rise in "reshoring"/"nearshoring". In the latter case, it has been observed that trade in intermediate goods may have weakened recently and could signal reshoring activity or at least a lack of further offshoring (HSBC Bank, 2015). However, views on reshoring/nearshoring remain inconclusive. Some observed trends suggest that a number of manufacturers are offshoring certain operations while at the same time bringing other activities back home or closer to home. Therefore, while some reshoring may be taking place due to increasing labour costs in the offshore locations, factors other than labour costs are taken into account when making relevant decisions about production sites. These include the quality of labour and access to foreign markets such as the Chinese markets (Cohen and Lee, 2015).

Table 1.2. Growth in the volume of merchandise, 2012–2014 (annual percentage change)

2012	2013	2014	Countries/regions	2012	2013	2014
2.0	2.6	2.3	**WORLD**	2.0	2.3	2.3
0.6	1.4	2.0	**Developed economies**	-0.4	-0.3	3.2
			of which:			
-0.1	1.7	1.5	European Union	-2.5	-0.9	2.8
-1.0	-1.9	0.6	Japan	3.8	0.5	2.8
3.9	2.6	3.1	United States	2.8	0.8	4.7
4.0	4.2	2.9	**Developing economies**	5.1	6.1	2.0
			of which:			
5.5	-2.0	-3.6	Africa	13.2	5.2	3.3
3.2	2.1	2.4	Developing America	3.3	4.0	0.6
4.0	5.2	3.5	Asia	7.7	6.6	2.2
			of which:			
6.2	7.7	6.8	China	3.6	9.9	3.9
-1.8	8.5	3.2	India	5.9	-0.2	3.2
9.6	3.1	0.3	Western Asia	9.2	9.6	0.2
0.7	1.8	0.2	**Transition economies**	5.6	-0.8	-8.5

Source: UNCTAD secretariat, based on UNCTADstat.
Note: Data on trade volumes are derived from international merchandise trade values deflated by UNCTAD unit value indices.

Consequently, it may be argued that long-term trade recovery depends on trends in GDP growth as well as on how the relationship between trade and GDP unfolds and whether relevant initiatives to further stimulate demand and trade are implemented. These may include stimulating demand for investment goods (for example, capital goods, transport and equipment) that are more import intensive; reorganizing supply chains with a new scope for the division of international labour, including in South Asia, sub-Saharan Africa and South America; increasing trade finance; furthering the liberalization of trade and reducing protective measures. In this respect, the potential for greater trade liberalization is firming up with the adoption of the World Trade Organization (WTO) Trade Facilitation Agreement (TFA) and the negotiations relating to the potential expansion of the WTO Information Technology Agreement. Other initiatives including, among others, the Transatlantic Trade and Investment Partnership between the European Union and the United States, which could raise the transatlantic annual GDP by $210 billion (Francois et al., 2013) and the Trans-Pacific Partnership, which could boost world income by $295 billion, also have the potential to further stimulate global trade (Petri and Plummer, 2012).

B. WORLD SEABORNE TRADE

1. General trends in seaborne trade

Although the responsiveness of trade to GDP growth may have moderated over recent years, demand for maritime transport services and seaborne trade volumes continue to be shaped by global economic growth and the need to carry merchandise trade. Figure 1.1 highlights the association between economic growth and industrial activity, as measured in this particular case by the Organization for Economic Cooperation and Development (OECD) Industrial Production Index, merchandise trade and seaborne shipments.

Preliminary estimates indicate that the volume of world seaborne shipments expanded by 3.4 per cent in 2014, that is, at the same rate as in 2013. Additions to volumes exceeded 300 million tons, taking the total to 9.84 billion, or around four fifths of total world merchandise trade. Dry cargo was estimated to have accounted for over two thirds of the total, while the share of tanker trade, including crude oil, petroleum products and gas was estimated to have slightly declined from

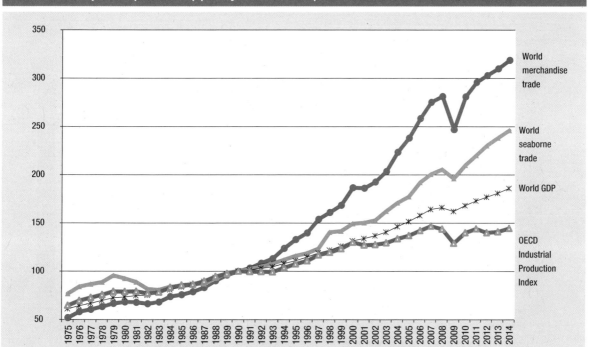

Figure 1.1. The OECD Industrial Production Index and indices for world GDP, merchandise trade and seaborne shipments (1975–2014) (base year 1990 = 100)

Sources: UNCTAD secretariat, based on OECD Main Economic Indicators, June 2015; United Nations Department of Economic and Social Affairs, 2015; LINK Global Economic Outlook, June 2015; UNCTAD Review of Maritime Transport, various issues; WTO, appendix table A1a, World merchandise exports, production and gross domestic product, 1950–2012; WTO press release 739, 14 April 2015.

Table 1.3. Developments in international seaborne trade, selected years (millions of tons loaded)

Year	Oil and gas	Main bulks[a]	Other dry cargo	Total (all cargoes)
1970	1 440	448	717	2 605
1980	1 871	608	1 225	3 704
1990	1 755	988	1 265	4 008
2000	2 163	1 295	2 526	5 984
2005	2 422	1 709	2 978	7 109
2006	2 698	1 814	3 188	7 700
2007	2 747	1 953	3 334	8 034
2008	2 742	2 065	3 422	8 229
2009	2 642	2 085	3 131	7 858
2010	2 772	2 335	3 302	8 409
2011	2 794	2 486	3 505	8 784
2012	2 841	2 742	3 614	9 197
2013	2 829	2 923	3 762	9 514
2014	2 826	3 112	3 903	9 842

Sources: UNCTAD secretariat, based on data supplied by reporting countries and as published on the relevant government and port industry websites, and by specialist sources. Data for 2006 onwards have been revised and updated to reflect improved reporting, including more recent figures and better information regarding the breakdown by cargo type. Figures for 2014 are estimated based on preliminary data or on the last year for which data were available.

[a] Iron ore, grain, coal, bauxite/alumina and phosphate rock; the data for 2006 onwards are based on various issues of the *Dry Bulk Trade Outlook*, produced by Clarksons Research.

Figure 1.2. International seaborne trade, selected years (millions of tons loaded)

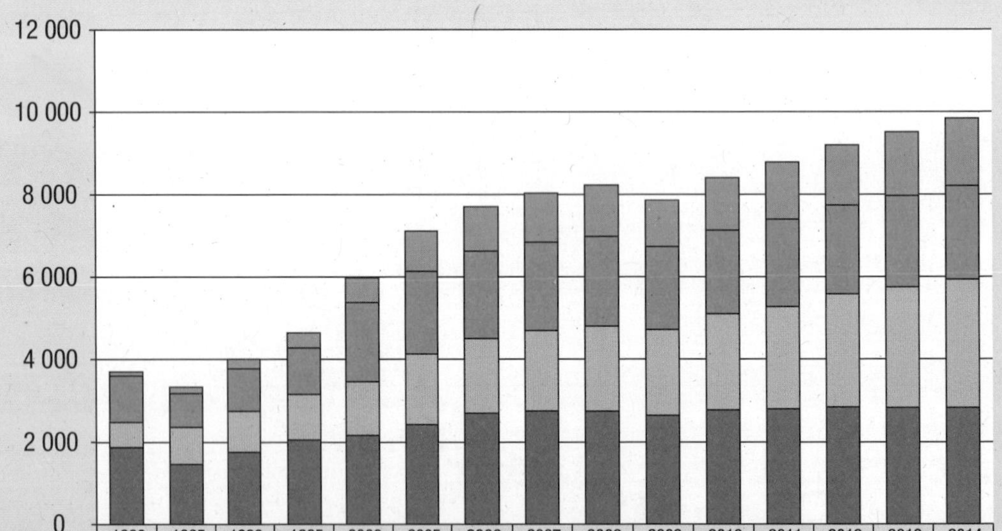

	1980	1985	1990	1995	2000	2005	2006	2007	2008	2009	2010	2011	2012	2013	2014
Container	102	152	234	371	598	969	1 076	1 193	1 249	1 127	1 280	1 393	1 464	1 544	1631
Other dry cargo	1 123	819	1 031	1 125	1 928	2 009	2 112	2 141	2 173	2 004	2 022	2 112	2 150	2 218	2272
Five major bulks	608	900	988	1 105	1 295	1 709	1 814	1 953	2 065	2 085	2 335	2 486	2 742	2 923	3112
Oil and gas	1 871	1 459	1 755	2 050	2 163	2 422	2 698	2 747	2 742	2 642	2 772	2 794	2 841	2 829	2 826

Sources: UNCTAD, *Review of Maritime Transport*, various issues. For 2006–2014, the breakdown by type of cargo is based on Clarksons Research, *Shipping Review and Outlook*, various issues.

nearly 30.0 per cent in 2013 to 28.7 per cent in 2014 (see tables 1.3, 1.4 (a), 1.4 (b) and figure 1.2).

Dry cargo shipments increased by 5.0 per cent, while tanker trade contracted by 1.6 per cent. Within dry cargo, dry bulk trade, including the five major bulk commodities (iron ore, coal, grain, bauxite/alumina and phosphate rock) as well as the minor bulk commodities (agribulks, metals and minerals, and manufactures) is estimated to have increased by 5.0 per cent, taking the total to 4.55 billion tons. Although growth in coal trade is estimated to have decelerated significantly to 2.8 per cent as compared with over 12.0 per cent in 2012 and 5.0 per cent in 2013, dry bulk shipments continued to be supported by the rapid expansion of global iron ore volumes, which in turn, was driven by China's continued strong import demand.

"Other dry cargo" (general cargo, break bulk and containerized) accounted for 35.2 per cent of all dry cargo shipments and is estimated to have increased by 4.9 per cent to reach 2.47 billion tons. Containerized trade, which accounted for about two thirds of "other dry cargo", was estimated to have increased by a strong 5.6 per cent, taking the total to 1.63 billion tons. In 2014, the performance of tanker trade weakened as compared with the previous year. Crude oil shipments contracted (-1.6 per cent), while petroleum products (+1.7 per cent) and gas trades (+3.9 per cent) expanded at slower rates. The structure of world seaborne trade is presented in figure 1.3.

Developing countries continued to contribute larger shares to international seaborne trade. Their contribution in terms of global goods loaded was estimated at 60 per cent, while their import demand as measured by the volume of goods unloaded reached 61 per cent (see figure 1.4 (a)). Behind the headline figures however, the individual contributions vary by regions and type of cargo, reflecting among other factors, differences in countries' economic structures, composition of trade, urbanization and level of development, as well as levels of integration into global trading networks and supply chains.

Over the past decade, developing countries have incrementally shifted patterns of trade. Since the 1970s, the distribution between the goods loaded and unloaded has changed significantly. As shown in figure 1.4 (b), over the years developing countries have become major importers and exporters and a driving force underpinning seaborne trade flows and

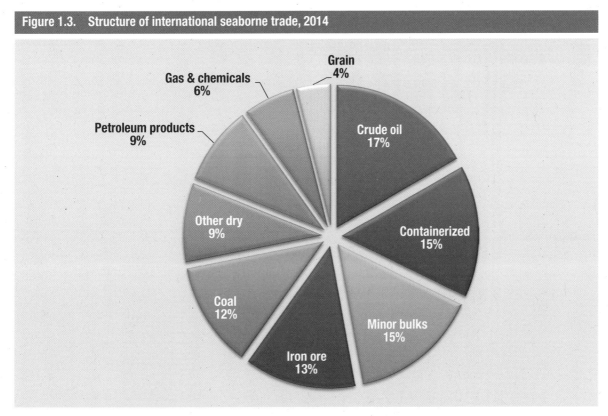

Figure 1.3. Structure of international seaborne trade, 2014

Source: UNCTAD secretariat, based on Clarksons Research, Seaborne Trade Monitor, 2(5), May 2015.

Table 1.4 (a). World seaborne trade 2006–2014, by type of cargo, country group and region (millions of tons)

Country group	Year	Goods loaded				Goods unloaded			
		Total	Crude	Petroleum products and gas	Dry cargo	Total	Crude	Petroleum products and gas	Dry cargo
		Millions of tons							
World	2006	7 700.3	1 783.4	914.8	5 002.1	7 878.3	1 931.2	893.7	5 053.4
	2007	8 034.1	1 813.4	933.5	5 287.1	8 140.2	1 995.7	903.8	5 240.8
	2008	8 229.5	1 785.2	957.0	5 487.2	8 286.3	1 942.3	934.9	5 409.2
	2009	7 858.0	1 710.5	931.1	5 216.4	7 832.0	1 874.1	921.3	5 036.6
	2010	8 408.9	1 787.7	983.8	5 637.5	8 443.8	1 933.2	979.2	5 531.4
	2011	8 784.3	1 759.5	1 034.2	5 990.5	8 797.7	1 896.5	1 037.7	5 863.5
	2012	9 196.7	1 785.7	1 055.0	6 356.0	9 188.5	1 929.5	1 055.1	6 203.8
	2013	9 513.6	1 737.9	1 090.8	6 684.8	9 500.1	1 882.0	1 095.2	6 523.0
	2014	9 841.7	1 710.3	1 116.1	7 015.3	9 808.4	1 861.5	1 122.6	6 824.2
Developed economies	2006	2 460.5	132.9	336.4	1 991.3	4 164.7	1 282.0	535.5	2 347.2
	2007	2 608.9	135.1	363.0	2 110.8	3 990.5	1 246.0	524.0	2 220.5
	2008	2 715.4	129.0	405.3	2 181.1	4 007.9	1 251.1	523.8	2 233.0
	2009	2 554.3	115.0	383.8	2 055.5	3 374.4	1 125.3	529.9	1 719.2
	2010	2 865.4	135.9	422.3	2 307.3	3 604.5	1 165.4	522.6	1 916.5
	2011	2 982.5	117.5	451.9	2 413.1	3 632.3	1 085.6	581.3	1 965.4
	2012	3 122.9	125.2	459.7	2 538.0	3 700.2	1 092.6	556.5	2 051.1
	2013	3 188.3	114.4	470.5	2 603.4	3 679.4	1 006.7	556.6	2 116.0
	2014	3 370.8	111.8	486.7	2 772.3	3 744.1	985.4	552.4	2 206.3
Transition economies	2006	410.3	123.1	41.3	245.9	70.6	5.6	3.1	61.9
	2007	407.9	124.4	39.9	243.7	76.8	7.3	3.5	66.0
	2008	431.5	138.2	36.7	256.6	89.3	6.3	3.8	79.2
	2009	505.3	142.1	44.4	318.8	93.3	3.5	4.6	85.3
	2010	515.7	150.2	45.9	319.7	122.1	3.5	4.6	114.0
	2011	505.0	132.6	42.0	330.5	156.7	4.2	4.4	148.1
	2012	544.2	135.6	40.3	368.3	148.1	3.8	4.0	140.3
	2013	551.9	145.1	32.1	374.8	77.4	1.1	10.6	65.7
	2014	591.2	136.1	43.4	411.8	80.1	0.9	11.2	67.9
Developing economies	2006	4 829.5	1 527.5	537.1	2 765.0	3 642.9	643.6	355.1	2 644.3
	2007	5 017.2	1 553.9	530.7	2 932.6	4 073.0	742.4	376.3	2 954.3
	2008	5 082.6	1 518.0	515.1	3 049.6	4 189.1	684.9	407.2	3 097.0
	2009	4 798.4	1 453.5	502.9	2 842.0	4 364.2	745.3	386.9	3 232.1
	2010	5 027.8	1 501.6	515.6	3 010.5	4 717.3	764.4	452.0	3 500.9
	2011	5 296.8	1 509.4	540.4	3 247.0	5 008.8	806.7	452.1	3 750.0
	2012	5 529.6	1 524.9	555.0	3 449.7	5 340.1	833.1	494.7	4 012.4
	2013	5 773.4	1 478.5	588.2	3 706.7	5 743.4	874.2	527.9	4 341.3
	2014	5 879.7	1 462.4	586.0	3 831.3	5 984.3	875.3	559.0	4 550.0

Table 1.4 (a). World seaborne trade 2006–2014, by type of cargo, country group and region (millions of tons) (continued)

Country group	Year	Goods loaded				Goods unloaded			
		Total	Crude	Petroleum products and gas	Dry cargo	Total	Crude	Petroleum products and gas	Dry cargo
		Millions of tons							
Africa	2006	721.9	353.8	86.0	282.2	349.8	41.3	39.4	269.1
	2007	732.0	362.5	81.8	287.6	380.0	45.7	44.5	289.8
	2008	766.7	379.2	83.3	304.2	376.6	45.0	43.5	288.1
	2009	708.0	354.0	83.0	271.0	386.8	44.6	39.7	302.5
	2010	754.0	351.1	92.0	310.9	416.9	42.7	40.5	333.7
	2011	723.7	338.0	68.5	317.2	378.2	37.8	46.3	294.1
	2012	757.8	364.2	70.2	323.4	393.6	32.8	51.0	309.8
	2013	815.3	327.5	82.4	405.3	432.2	36.6	65.3	330.3
	2014	761.3	301.4	78.3	381.6	466.0	36.4	69.3	360.3
America	2006	1 030.7	251.3	93.9	685.5	373.4	49.6	60.1	263.7
	2007	1 067.1	252.3	90.7	724.2	415.9	76.0	64.0	275.9
	2008	1 108.2	234.6	93.0	780.6	436.8	74.2	69.9	292.7
	2009	1 029.8	225.7	74.0	730.1	371.9	64.4	73.6	234.0
	2010	1 172.6	241.6	85.1	846.0	448.7	69.9	74.7	304.2
	2011	1 239.2	253.8	83.5	901.9	508.3	71.1	73.9	363.4
	2012	1 282.6	253.3	85.9	943.4	546.7	74.6	83.6	388.5
	2013	1 263.7	240.0	69.8	953.9	569.4	69.4	89.4	410.7
	2014	1 283.6	232.0	72.6	979.0	606.9	70.0	92.7	444.3
Asia	2006	3 073.1	921.2	357.0	1 794.8	2 906.8	552.7	248.8	2 105.3
	2007	3 214.6	938.2	358.1	1 918.3	3 263.6	620.7	260.8	2 382.1
	2008	3 203.6	902.7	338.6	1 962.2	3 361.9	565.6	286.8	2 509.5
	2009	3 054.3	872.3	345.8	1 836.3	3 592.4	636.3	269.9	2 686.2
	2010	3 094.6	907.5	338.3	1 848.8	3 838.2	651.8	333.1	2 853.4
	2011	3 326.7	916.0	388.2	2 022.6	4 108.8	697.8	328.0	3 082.9
	2012	3 480.9	905.8	398.1	2 177.0	4 386.9	725.7	355.5	3 305.7
	2013	3 686.9	909.4	435.2	2 342.4	4 728.7	767.4	369.2	3 592.1
	2014	3 826.8	927.3	434.3	2 465.2	4 897.2	768.0	392.6	3 736.5
Oceania	2006	3.8	1.2	0.1	2.5	12.9	0.0	6.7	6.2
	2007	3.5	0.9	0.1	2.5	13.5	0.0	7.0	6.5
	2008	4.2	1.5	0.1	2.6	13.8	0.0	7.1	6.7
	2009	6.3	1.5	0.2	4.6	13.1	0.0	3.6	9.5
	2010	6.5	1.5	0.2	4.8	13.4	0.0	3.7	9.7
	2011	7.1	1.6	0.2	5.3	13.5	0.0	3.9	9.6
	2012	8.3	1.6	0.8	5.9	13.0	0.0	4.6	8.4
	2013	7.5	1.6	0.8	5.1	13.1	0.8	4.1	8.2
	2014	8.1	1.6	0.9	5.5	14.2	0.9	4.4	8.9

Table 1.4 (b). World seaborne trade 2006–2014, by type of cargo, country group and region (percentage share)

Country group	Year	Goods loaded				Goods unloaded			
		Total	Crude	Petroleum products and gas	Dry cargo	Total	Crude	Petroleum products and gas	Dry cargo
		Percentage share							
World	2006	100.0	23.2	11.9	65.0	100.0	24.5	11.3	64.1
	2007	100.0	22.6	11.6	65.8	100.0	24.5	11.1	64.4
	2008	100.0	21.7	11.6	66.7	100.0	23.4	11.3	65.3
	2009	100.0	21.8	11.8	66.4	100.0	23.9	11.8	64.3
	2010	100.0	21.3	11.7	67.0	100.0	22.9	11.6	65.5
	2011	100.0	20.0	11.8	68.2	100.0	21.6	11.8	66.6
	2012	100.0	19.4	11.5	69.1	100.0	21.0	11.5	67.5
	2013	100.0	18.3	11.5	70.3	100.0	19.8	11.5	68.7
	2014	100.0	17.4	11.3	71.3	100.0	19.0	11.4	69.6
Developed economies	2006	32.0	7.4	36.8	39.8	52.9	66.4	59.9	46.4
	2007	32.5	7.5	38.9	39.9	49.0	62.4	58.0	42.4
	2008	33.0	7.2	42.3	39.7	48.4	64.4	56.0	41.3
	2009	32.5	6.7	41.2	39.4	43.1	60.0	57.5	34.1
	2010	34.1	7.6	42.9	40.9	42.7	60.3	53.4	34.6
	2011	34.0	6.7	43.7	40.3	41.3	57.2	56.0	33.5
	2012	34.0	7.0	43.6	39.9	40.3	56.6	52.7	33.1
	2013	33.5	6.6	43.1	38.9	38.7	53.5	50.8	32.4
	2014	34.3	6.5	43.6	39.5	38.2	52.9	49.2	32.3
Transition economies	2006	5.3	6.9	4.5	4.9	0.9	0.3	0.3	1.2
	2007	5.1	6.9	4.3	4.6	0.9	0.4	0.4	1.3
	2008	5.2	7.7	3.8	4.7	1.1	0.3	0.4	1.5
	2009	6.4	8.3	4.8	6.1	1.2	0.2	0.5	1.7
	2010	6.1	8.4	4.7	5.7	1.4	0.2	0.5	2.1
	2011	5.7	7.5	4.1	5.5	1.8	0.2	0.4	2.5
	2012	5.9	7.6	3.8	5.8	1.6	0.2	0.4	2.3
	2013	5.8	8.3	2.9	5.6	0.8	0.1	1.0	1.0
	2014	6.0	8.0	3.9	5.9	0.8	0.0	1.0	1.0
Developing economies	2006	62.7	85.6	58.7	55.3	46.2	33.3	39.7	52.3
	2007	62.4	85.7	56.9	55.5	50.0	37.2	41.6	56.4
	2008	61.8	85.0	53.8	55.6	50.6	35.3	43.6	57.3
	2009	61.1	85.0	54.0	54.5	55.7	39.8	42.0	64.2
	2010	59.8	84.0	52.4	53.4	55.9	39.5	46.2	63.3
	2011	60.3	85.8	52.2	54.2	56.9	42.5	43.6	64.0
	2012	60.1	85.4	52.6	54.3	58.1	43.2	46.9	64.7
	2013	60.7	85.1	53.9	55.4	60.5	46.4	48.2	66.6
	2014	59.7	85.5	52.5	54.6	61.0	47.0	49.8	66.7
Africa	2006	9.4	19.8	9.4	5.6	4.4	2.1	4.4	5.3
	2007	9.1	20.0	8.8	5.4	4.7	2.3	4.9	5.5
	2008	9.3	21.2	8.7	5.5	4.5	2.3	4.7	5.3
	2009	9.0	20.7	8.9	5.2	4.9	2.4	4.3	6.0
	2010	9.0	19.6	9.4	5.5	4.9	2.2	4.1	6.0
	2011	8.2	19.2	6.6	5.3	4.3	2.0	4.5	5.0
	2012	8.2	20.4	6.6	5.1	4.3	1.7	4.8	5.0
	2013	8.6	18.8	7.6	6.1	4.5	1.9	6.0	5.1
	2014	7.7	17.6	7.0	5.4	4.8	2.0	6.2	5.3

CHAPTER 1: DEVELOPMENTS IN INTERNATIONAL SEABORNE TRADE

Table 1.4 (b). World seaborne trade 2006–2014, by type of cargo, country group and region (percentage share) (continued)

Country group	Year	Goods loaded				Goods unloaded			
		Total	Crude	Petroleum products and gas	Dry cargo	Total	Crude	Petroleum products and gas	Dry cargo
		Percentage share							
America	2006	13.4	14.1	10.3	13.7	4.7	2.6	6.7	5.2
	2007	13.3	13.9	9.7	13.7	5.1	3.8	7.1	5.3
	2008	13.5	13.1	9.7	14.2	5.3	3.8	7.5	5.4
	2009	13.1	13.2	7.9	14.0	4.7	3.4	8.0	4.6
	2010	13.9	13.5	8.7	15.0	5.3	3.6	7.6	5.5
	2011	14.1	14.4	8.1	15.1	5.8	3.7	7.1	6.2
	2012	13.9	14.2	8.1	14.8	5.9	3.9	7.9	6.3
	2013	13.3	13.8	6.4	14.3	6.0	3.7	8.2	6.3
	2014	13.0	13.6	6.5	14.0	6.2	3.8	8.3	6.5
Asia	2006	39.9	51.7	39.0	35.9	36.9	28.6	27.8	41.7
	2007	40.0	51.7	38.4	36.3	40.1	31.1	28.9	45.5
	2008	38.9	50.6	35.4	35.8	40.6	29.1	30.7	46.4
	2009	38.9	51.0	37.1	35.2	45.9	34.0	29.3	53.3
	2010	36.8	50.8	34.4	32.8	45.5	33.7	34.0	51.6
	2011	37.9	52.1	37.5	33.8	46.7	36.8	31.6	52.6
	2012	37.8	50.7	37.7	34.3	47.7	37.6	33.7	53.3
	2013	38.8	52.3	39.9	35.0	49.8	40.8	33.7	55.1
	2014	38.9	54.2	38.9	35.1	49.9	41.3	35.0	54.8
Oceania	2006	0.0	0.1	0.01	0.0	0.2	-	0.7	0.1
	2007	0.0	0.1	0.01	0.0	0.2	-	0.8	0.1
	2008	0.1	0.1	0.01	0.0	0.2	-	0.8	0.1
	2009	0.1	0.1	0.02	0.1	0.2	-	0.4	0.2
	2010	0.1	0.1	0.0	0.1	0.2	-	0.4	0.2
	2011	0.1	0.1	0.0	0.1	0.2	-	0.4	0.2
	2012	0.1	0.1	0.1	0.1	0.1	-	0.4	0.1
	2013	0.1	0.1	0.1	0.1	0.1	-	0.4	0.1
	2014	0.1	0.1	0.1	0.1	0.1	-	0.4	0.1

Sources: UNCTAD secretariat, based on data supplied by reporting countries and as published on the relevant government and port industry websites, and by specialist sources. Data from 2006 onwards have been revised and updated to reflect improved reporting, including more recent figures and better information regarding the breakdown by cargo type. Figures for 2014 are estimated based on preliminary data or on the last year for which data were available.

demand for maritime transport services. They are no longer only sources of supply of raw materials, but also key players in globalized manufacturing processes and a growing source of demand. In terms of regional influence, Asia continued to dominate as the main loading and unloading area in 2014, followed by the Americas, Europe, Oceania and Africa (figure 1.4 (c)).

The impact of the drop in oil price levels since June 2014 extends beyond the energy markets and the world economy to also affect shipping and seaborne trade, in particular tanker trade. Indirect impacts are felt through changes in the areas of activity and sectors that generate the demand for maritime transport services. These include changes in production costs, economic growth, income and purchasing power of oil producers/exporters and consumers/importers, terms of trade, and investments in oil and gas, as well as investments in alternative fuels and fuel efficient technologies. Meanwhile, direct impacts on shipping and seaborne trade are reflected in lower fuel and transport costs. Ship bunker fuel costs have fallen significantly over the past few months. For example, the 380 centistoke bunker prices in Rotterdam dropped from $590 per ton in June 2014 to $318 per ton in December 2014, a drop of 46 per cent (Clarksons Research, 2015a). Lower fuel costs reduce ship operators' expenditure and rates paid by shippers. This, in turn, can stimulate the demand for maritime transport services and increase seaborne cargo flows.

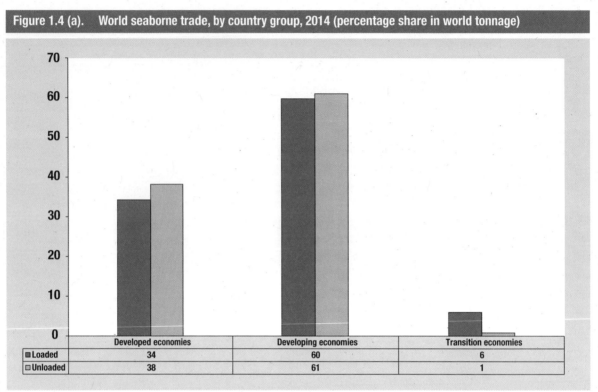

Figure 1.4 (a). World seaborne trade, by country group, 2014 (percentage share in world tonnage)

	Developed economies	Developing economies	Transition economies
Loaded	34	60	6
Unloaded	38	61	1

Sources: UNCTAD secretariat, based on data supplied by reporting countries and as published on the relevant government and port industry websites, and by specialist sources. Estimated figures are based on preliminary data or on the last year for which data were available.

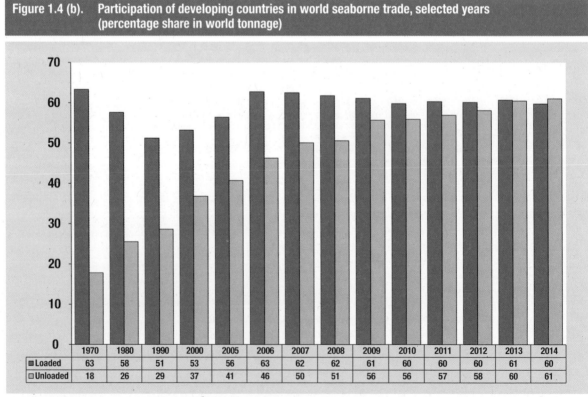

Figure 1.4 (b). Participation of developing countries in world seaborne trade, selected years (percentage share in world tonnage)

	1970	1980	1990	2000	2005	2006	2007	2008	2009	2010	2011	2012	2013	2014
Loaded	63	58	51	53	56	63	62	62	61	60	60	60	61	60
Unloaded	18	26	29	37	41	46	50	51	56	56	57	58	60	61

Source: UNCTAD *Review of Maritime Transport*, various issues.

CHAPTER 1: DEVELOPMENTS IN INTERNATIONAL SEABORNE TRADE

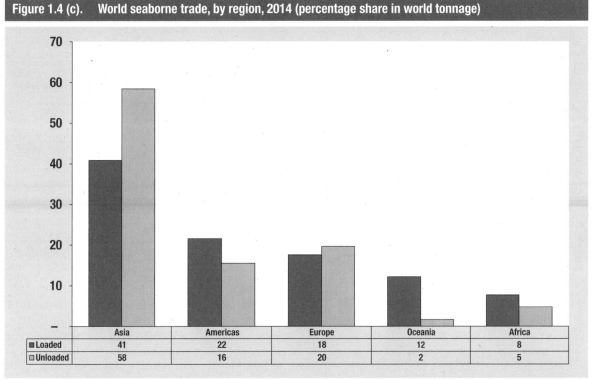

Figure 1.4 (c). World seaborne trade, by region, 2014 (percentage share in world tonnage)

	Asia	Americas	Europe	Oceania	Africa
Loaded	41	22	18	12	8
Unloaded	58	16	20	2	5

Sources: UNCTAD secretariat, based on data supplied by reporting countries and as published on the relevant government and port industry websites, and by specialist sources. Estimated figures are based on preliminary data or on the last year for which data were available.

In addition to being potentially beneficial to shippers and trade generally, it may be argued that lower bunker fuel costs can further shape the global shipping networks and enhance market access and connectivity by making, for example, additional port calls on existing services more cost-effective. Furthermore, in addition to supporting demand and therefore larger crude trade volumes, lower oil prices and the related "contango" can lead to the use of tankers as storage units to store oil. Although a number of fixtures were reported in 2014 and early 2015, oil storage did not become as widespread as initially expected given the less promising trends in oil futures and the rise in charter rates (Clarksons Research, 2015a).

Some observers have commented that a lower price and cost environment could potentially undermine the competitiveness of energy-efficient ships and "eco-ship" designs and equipment (*Ship & Bunker*, 2014a). Others have argued that the benefits generated from slow steaming, a major cost-cutting measure implemented since 2008/2009, could be eroded as ships resume sailing at faster speeds (*Journal of Commerce (JOC)*, 2014). While uncertainty about the future of slow steaming remains, so far it would appear that average operating speeds have not increased, owing probably to the slower design speed of eco-ships and the risk for profitability. Faster speeds are likely to liberate excess capacity back into some shipping markets and therefore undermine the fundamentals of the market and the profitability (*Lloyd's List*, 2015a). It was noted that if carriers were, for example, to speed up their services to remove one week from transit times on the Asia–Europe container route, they would be adding 2.5 per cent to the existing capacity on the route (*Lloyd's List*, 2015b). To put this in perspective and based on information obtained from Clarksons Research, it should be noted that prior to implementing slow steaming, a typical structure for a journey from the Far East to Europe, for example, included eight ship services to maintain weekly calls over a period of 56 days for full rotation (28 days for one leg). With the implementation of slow steaming, the number of ship services increased to ten to maintain weekly calls, while transit times increased to 70 days for a full rotation (35 days for one leg).

A related development that affects the shipping industry is the coming into force on 1 January 2015 of the requirement under the International Convention for the Prevention of Pollution from Ships, 1973, as modified by the Protocol of 1978 (MARPOL) annex

VI (Regulations for the Prevention of Air Pollution from Ships), specifically under regulation 14, which covers emissions of sulphur oxides (SO_x) and particulate matter from ships. The ECAs were established under MARPOL annex VI for SO_x and include the Baltic Sea area, the North Sea area, the North American Atlantic area, and the United States Caribbean Sea area. Ships trading in ECAs are required to use fuel oil with a sulphur content of no more than 0.10 per cent from 1 January 2015. The previous limit was 1.00 per cent (IMO, 2015). The current limit applied in waters other than ECAs is 3.50 per cent and is set to drop to 0.50 per cent on and after 1 January 2020; however, the coming into force of this latter limit is subject to a review to be completed by 2018 regarding the availability of the required fuel oil (IMO, 2015). Although ship operators were concerned about the cost of using more expensive lower sulphur fuels, the lower oil price environment has helped offset the price premium, with the cost of cleaner fuel remaining reasonable given the general lower oil prices and bunker fuel costs (Barnard, 2015). However, in anticipation of the potential increase in bunker fuel costs, some carriers have announced some surcharges that will be applied if necessary.

2. Seaborne trade in ton–miles

The ton–mile unit offers a more accurate measure of demand for shipping services and tonnage as it takes into account distance, which determines ships' transportation capacity over time. In 2014, growth in ton–miles performed by maritime transportation was estimated to have increased by 4.4 per cent, up from 3.1 per cent in 2013 (see figure 1.5) (Clarksons Research, 2015b). Dry bulk commodities, namely iron ore, coal, grain, bauxite and alumina, phosphate rock and minor bulks accounted for nearly half of the total 52,572 estimated billion ton—miles performed in 2014. The ton—miles of the dry bulks expanded at a firm rate (6.4 per cent for major dry bulk commodities and 5.2 per cent for minor bulks). Ton-miles generated by containerized trade were estimated to have increased by 5.4 per cent (Clarksons Research, 2015b), driven by the recovery on the peak legs of the Asia–Europe and trans-Pacific trade routes as well as the continued rise in the longer haul North–South trade volumes. Coal and iron ore import demand from Asia has contributed significantly to the growth in dry bulk trade volumes over recent years. Apart from China, iron ore and coal demand from other fast growing economies such as India and the Republic of Korea has also been on the rise.

With crude oil volumes estimated to have contracted in 2014, the associated ton–miles remained flat, indicating growth in distances travelled. The average haul of crude oil trade to Asia was estimated at over 5,000 miles in 2014, or 9 per cent greater than 2005 levels (Elliott-Green, 2015). China has been driving growth given its increasing sourcing of crude oil imports from various locations, including both long and shorter haul routes (for example, the Caribbean, West Africa, Western Asia and the Russian Federation). India is also increasingly sourcing crude oil imports from Western Asia, Western Africa and the Caribbean, resulting in growing long-haul imports. The average haul of Indian crude oil imports was estimated at over 4,000 miles in 2014, up from 1,900 miles in 2005 (Elliott-Green, 2015). The United States has also contributed to the ton–mile trends observed over recent years. While its crude oil imports have fallen by nearly half since 2005, its crude oil ton–miles have declined less rapidly. This reflects the United States' oil trade patterns as larger import declines were recorded on the short-haul trades (for example, West Africa) as opposed to the longer haul Western Asian route. In 2014, the average haul of crude oil imports into the United States increased to 7,000 miles, representing an 18 per cent increase over 2005 (Elliott-Green, 2015). Ton–miles generated by the trade of petroleum products increased by 3.8 per cent, while gas trade ton–miles expanded by 2.6 per cent, driven mainly by growth in the liquefied petroleum gas (LPG) flows (Clarksons Research, 2015b).

3. Seaborne trade by cargo type

(a) Tanker trade

Crude oil

While oil prices are an important market signal, other factors are also increasingly shaping the tanker trade landscape. These include the response of shale oil producers to the lower oil price levels, policy decisions by members of the Organization of the Petroleum Exporting Countries, geopolitical developments, and political tensions.

Reflecting subdued growth in global oil consumption in 2014 (+0.8 per cent) (International Energy Agency, 2015), crude oil shipments were estimated at 1.7 billion tons in 2014, a drop of 1.7 per cent over the previous year. The firm import demand of Asian countries, in particular China and India, the effect of lower oil prices on stock building, and increased oil

CHAPTER 1: DEVELOPMENTS IN INTERNATIONAL SEABORNE TRADE

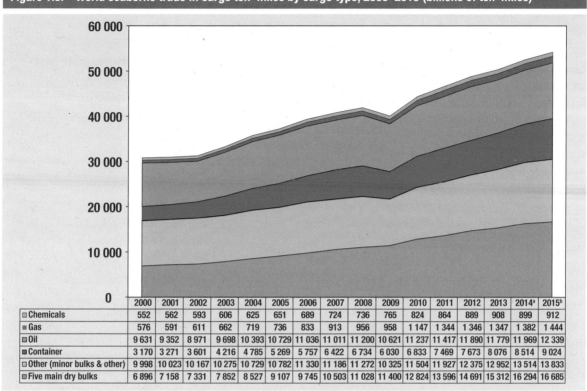

Figure 1.5. World seaborne trade in cargo ton–miles by cargo type, 2000–2015 (billions of ton–miles)

	2000	2001	2002	2003	2004	2005	2006	2007	2008	2009	2010	2011	2012	2013	2014a	2015b
Chemicals	552	562	593	606	625	651	689	724	736	765	824	864	889	908	899	912
Gas	576	591	611	662	719	736	833	913	956	958	1 147	1 344	1 346	1 347	1 382	1 444
Oil	9 631	9 352	8 971	9 698	10 393	10 729	11 036	11 011	11 200	10 621	11 237	11 417	11 890	11 779	11 969	12 339
Container	3 170	3 271	3 601	4 216	4 785	5 269	5 757	6 422	6 734	6 030	6 833	7 469	7 673	8 076	8 514	9 024
Other (minor bulks & other)	9 998	10 023	10 167	10 275	10 729	10 782	11 330	11 186	11 272	10 325	11 504	11 927	12 375	12 952	13 514	13 833
Five main dry bulks	6 896	7 158	7 331	7 852	8 527	9 107	9 745	10 503	11 028	11 400	12 824	13 596	14 691	15 312	16 294	16 685

Source: UNCTAD secretariat, based on data from Clarksons Research (2015b).
a Estimated
b Forecast

supply (+2.5 per cent) have combined to offset the limited growth elsewhere and the decline in import volumes of the United States and Europe.

In 2014, crude oil imports into the United States declined by nearly 12 per cent to reach 4.5 million barrels per day, while imports into China increased by 9.8 per cent (5.6 million barrels per day) (Clarksons Research, 2015c) in tandem with its growing refinery capacity, strategic petroleum reserves requirements as well as the supporting effect of lower oil prices. This trend is likely to continue given the expected further growth in China's refinery capacity and petroleum reserve requirements. Underpinned by a rising national refinery capacity, India has over recent years emerged as an important crude oil importer (Clarksons Research, 2015d). On the export side, members of the Organization of the Petroleum Exporting Countries maintained the production levels to retain market share. African crude exports contracted by 4.6 per cent due to technical problems in Angola, infrastructure-related disruptions in Nigeria as well as conflicts in Libya. An overview of global consumers and producers of oil and gas is presented in table 1.5.

Table 1.5. Major producers and consumers of oil and natural gas, 2014 (world market share in percentage)

World oil production		World oil consumption	
Western Asia	32	Asia Pacific	34
North America	18	North America	22
Transition economies	16	Europe	15
Developing America	12	Developing America	10
Africa	9	Western Asia	9
Asia Pacific	9	Transition economies	5
Europe	3	Africa	4
World natural gas production		**World natural gas consumption**	
North America	26	North America	26
Transition economies	22	Asia Pacific	20
Western Asia	17	Transition economies	17
Asia Pacific	15	Western Asia	14
Europe	7	Europe	13
Developing America	7	Developing America	8
Africa	6	Africa	4

Source: UNCTAD secretariat on the basis of data published in the British Petroleum (BP) *Statistical Review of World Energy 2015* (June 2015).

Note: Oil includes crude oil, shale oil, oil sands and natural gas liquids NGLs – the liquid content of natural gas where this is recovered separately). The term excludes liquid fuels from other sources as biomass and coal derivatives.

Refined petroleum products

Developments in refinery capacities can significantly shape crude and product trade patterns. In 2014, the global refinery capacity increased by 1.4 per cent (British Petroleum, 2015), driven mainly by growth in Brazil, China, Singapore and Western Asia. According to UNCTAD's estimates, which include gas trade, the volume of petroleum products and gas loaded in 2014 increased by 2.3 per cent and reached 1.11 billion tons. Meanwhile, data from Clarksons Research indicate that petroleum products are estimated to have increased by 1.7 per cent in 2014 and reached 977 million tons, while gas trade increased by 3.9 per cent and totalled 319 million tons (Clarksons Research, 2015b).

On the supply side, increasing exports from Western Asia (+6.3 per cent), the United States (+4.0 per cent) and the economies in transition (+3.6 per cent) helped support growth (Clarksons Research, 2015b). Imports into Latin America (+11.8 per cent) and developing Asia (other than China) (+6.3 per cent) have been the main driver of growth. Meanwhile, imports into Africa, Australia, India, Japan and the Republic of Korea are estimated to have remained steady, while imports into China, the United States and Europe declined by 25 per cent, 12.5 per cent and 1.5 per cent, respectively (Clarksons Research, 2015b).

During recent years, China has moved away from being a net importer of oil products. Together, China's domestic oversupply of petroleum products, growing refinery capacity and reduced national demand have contributed to reducing import needs and increasing exports. Refinery capacity in Western Asia has also been on the rise, reflecting growing domestic requirements as well as export needs. Although the capacity growth was limited in the United States, throughput increased by 3.5 per cent, taking the country's global share to over 20 per cent in 2014 (British Petroleum, 2015).

Natural gas and liquefied gases

Liquefied natural gas (LNG) increased its share of global gas trade carried by sea in 2014. Volumes increased by 2.5 per cent, taking the total to 333.3 billion cubic metres. Growth was driven by higher import demand in China, India, the United Kingdom, Brazil and Mexico. Japan, the largest world importer, increased imports by 1.4 per cent, while the Republic of Korea, the second largest importer, recorded a decline of 5.7 per cent as inventory restocking was completed (British Petroleum, 2015). Rising import demand in developing Asia and America was supported by growing power generation, petrochemical and heating demand, as well as expanded regasification capacity in China and India.

Major exporters, including Qatar, reduced exports, while others such as Algeria, Australia, Malaysia and Papua New Guinea recorded increases in export volumes. Meanwhile, LNG imports into the United States have been curtailed by the shale revolution. However, the country has the potential to eventually emerge as an important gas exporter (British Petroleum, 2015).

Overall, firm global demand for LNG, led by the Asian economies, is expected to support growth in LNG carrier demand, while environmental regulations and air emission controls may lead to a growing role for gas. Some observers predict that LNG volumes will double by 2020, with Australia emerging as a world leading exporter together with other producers such as the Russian Federation, the United States, Canada and East Africa (*Lloyd's List*, 2015c). These developments will affect demand for gas carriers and further shape LNG trade flows and patterns.

Global LPG trade is estimated to have increased by 12.7 per cent in 2014 to reach 71 million tons. Growth was largely supported by the expansion of shale production in the United States and LPG exports (Clarksons Research, 2015a). Imports of LPG into China and India remained firm and contributed to raising long-haul trades and helping absorb more gas carrier capacity (Clarksons Research, 2015a).

(b) Dry cargo trade: Major and minor dry bulks and other dry cargo

The import demand of emerging developing economies, in particular China and India, remained the main driver of growth in dry bulk cargo shipments in 2014. During the year, the increase in world seaborne dry bulk shipments was estimated at 5.0 per cent, a slower rate than the previous four years (*Dry Bulk Trade Outlook*, 2015a). Growth was underpinned by the strong expansion in iron ore trade (+12.4 per cent) which accounted for about 30.0 per cent of all dry bulk cargo and reached 1.34 billion tons. In contrast, coal trade shipments were estimated to have increased by a modest 2.8 per cent, a much slower rate than the double-digit growth recorded in 2012

(+12.3 per cent). Shipments of the five major bulk commodities increased by 6.5 per cent, while the volume of minor bulk commodities is estimated to have increased by 2.0 per cent, reaching 3.1 billion tons and 1.43 billion tons, respectively. Exports of dry bulk commodities such as bauxite, nickel ore, iron ore and coal were constrained by, among other factors, bans on mining activities, restrictions on exports, weather patterns, regulatory measures and policies seeking to promote national producers and industries. An overview of global producers and users of steel as well as importers of select major dry bulk commodities is presented in table 1.6.

Iron ore shipments

Supported by increased production and exports from Australia, seaborne iron ore trade is estimated to have grown by 12.4 per cent, taking the total to 1.34 billion tons in 2014 (*Dry Bulk Trade Outlook*, 2015a). While growth in China's steel production decelerated in 2014 (World Steel Association, 2015), its iron ore imports remained robust due to lower international iron ore prices and the ample supply from Australia. The cheaper and higher quality imported iron ore displaced domestic supply. There are significant concerns, however, about the long-term developments in China's steel industry and related implications for dry bulk shipping. On the positive side for shipping, the increased Indian import demand may indicate the potential of India to further rely on iron ore imports to support its growing steel production sector. India's iron ore imports are currently expected to grow by 23 per cent in 2015.

Shipments from Australia are estimated to have increased by 24.2 per cent and accounted for over half of global iron ore exports in 2014. Exports from Brazil, which accounted for 25.3 per cent of world iron ore shipments, increased by 5.4 per cent. Exports from Sierra Leone grew by approximately 51.0 per cent to reach 18.1 million tons despite the negative impact of the Ebola outbreak on mining activities (*Dry Bulk Trade Outlook*, 2015b).

Looking forward, while, in the short term, iron ore shipments are expected to continue to grow, concerns relating to a slowdown of China's steel industry and import demand are causing uncertainty in the outlook for bulk carrier demand. Additionally, while lower iron ore prices stimulated iron ore trade in 2014, the sharp fall in prices raises concerns about the ability of some miners to continue production at a loss (Trimmel, 2015).

Table 1.6. Some major dry bulks and steel: Main producers, users, exporters and importers, 2014 (world market shares in percentages)

Steel producers		Steel users	
China	50	China	46
Japan	7	United States	7
United States	7	India	5
India	5	Japan	4
Republic of Korea	4	Republic of Korea	4
Russian Federation	4	Russian Federation	3
Germany	3	Transition economies	3
Turkey	2	Germany	3
Brazil	2	Turkey	2
Ukraine	2	Mexico	1
Other	15	Other	22
Iron ore exporters		*Iron ore importers*	
Australia	54	China	68
Brazil	25	Japan	10
South Africa	5	Europe	9
Canada	3	Republic of Korea	6
Sweden	2	Other	7
Other	12		
Coal exporters		*Coal importers*	
Indonesia	34	China	20
Australia	31	Europe	19
Russian Federation	9	India	18
Colombia	6	Japan	15
South Africa	6	Republic of Korea	11
Canada	3	Taiwan Province of China	5
Other	12	Malaysia	2
		Thailand	2
		Other	9
Grain exporters		*Grain importers*	
United States	26	Asia	33
European Union	14	Africa	21
Ukraine	10	Developing America	20
Canada	9	Western Asia	19
Argentina	8	Europe	5
Russian Federation	8	Transition economies	2
Others	25		

Sources: UNCTAD secretariat, based on data from World Steel Association, 2015; *Dry Bulk Trade Outlook* (May 2015a); Clarksons Research (2015b); and International Grains Council, Grains Market Report, June 2015.

Coal shipments

Growth in world coal shipments (thermal and coking) decelerated to 2.8 per cent with total volumes estimated at 1.2 billion tons. Thermal coal exports, which accounted for over two thirds of coal trade in 2014, are estimated to have increased by 3.8 per cent and reached 950 million tons. Coking coal shipments fell marginally (−0.8 per cent) to 262 million tons, owing mainly to reduced import demand from China (*Dry Bulk Trade Outlook*, 2015a).

China was the main engine fuelling the rapid expansion of world seaborne coal trade over the past decade, with its share of global coal shipments reaching 20.0 per cent in 2014, up from 2.0 per cent in 2005. An estimated 10.0 per cent drop in China's coal imports in 2014 may have a significant impact on dry bulk shipping demand. Factors contributing to the drop in China's imports include, among others, the falling import demand, which reflects China's regulations on saleable coal use, a slowdown in steel production, coal import taxes and quality limits, efforts to protect the domestic coal mining industry, hydroelectric power production and government initiatives to reduce air pollution.

Elsewhere, imports into the European Union have also dropped and are expected to further depress as member States comply with the Large Combustion Plant Directive (European Commission, 2001). The Directive contributed to reducing coal emissions by 5.0 per cent between 2008 and 2013, as some stations have already been closed (Jones and Worthington, 2014). Reflecting its growing steel production, India's coking coal imports are estimated to have grown by 24.3 per cent, while thermal coal imports grew by 7.1 per cent. On the export side, total thermal coal exports from Indonesia dropped by 1.7 per cent, while exports from the United States fell by 33.7 per cent, owing in particular to rising mining production costs, lower international coal prices and, generally, weaker global demand. Coking coals exports from the main exporters, including Canada, the Russian Federation and the United States, also declined in 2014, with the exception of exports from Australia (+3.6 per cent) (*Dry Bulk Trade Outlook*, 2015a).

Grain shipments

Reflecting improved weather conditions and harvest recovery in key exporters including Canada, the European Union, Ukraine and the United States, and, in the case of the Russian Federation, a favourable exchange rate, global grain shipments (including wheat, coarse grain and soybean) are estimated to have increased by 11.1 per cent in 2014 and totalled 430 million tons (*Dry Bulk Trade Outlook*, 2015a). Other exporters including Australia and Argentina recorded flat growth rates or contractions in export volumes during the crop year 2013/2014.

Japan, the top world importer, imported less grain (-1.3 per cent), while China, the second world importer, increased its imports, in particular of soybeans (+16.4 per cent). The strong demand from China will continue to support soybean export shipments from developing America. Other grain importers such as Algeria, Indonesia, the Islamic Republic of Iran, Mexico and Saudi Arabia increased their imports, while the economies in transition Brazil, Colombia, Morocco and Tunisia reduced their imports, given ample domestic supply.

Bauxite, alumina and phosphate rock

Bauxite trade continues to face uncertainty due to Indonesia's export restrictions, introduced in January 2014. Global bauxite and alumina trade volumes are estimated to have contracted by 24.5 per cent in 2014 to reach 105 million tons. China's import volumes of bauxite contracted by over half in 2014, a stark contrast with the 79.0 per cent increase of 2013, when refiners stockpiled the mineral in anticipation of the export ban (*Dry Bulk Trade Outlook*, 2015a). Indonesia used to be the largest exporter of bauxite to China. However, with the application of the export restrictions, China is increasingly sourcing its imports from Malaysia. In the meantime, Australia has the potential to emerge as an important supplier.

In 2014, global shipments of phosphate rock are estimated to have increased by 7.2 per cent, taking the total volume to 30 million tons. World phosphate rock production declined by 2.2 per cent, with contractions in output in China and the United States being to some extent offset by increased production in Morocco. Global production capacity is projected to grow owing to expansions in existing mines in Jordan, Kazakhstan, Morocco, Peru, the Russian Federation and Tunisia. World consumption of phosphorus pentoxide (P_2O_5) from phosphate rock is also projected to increase, with the largest growth occurring in Asia and developing America. These trends are likely to drive up shipments of phosphate rock and shape the associated flows and trade patterns.

Dry cargo: Minor bulks

Growth in global shipments of minor bulk commodities are estimated to have decelerated to 1.8 per cent in

2014, with total volumes reaching 1.43 billion tons. Manufactures (steel and forest products) accounted for 41.9 per cent of the total followed by metals and minerals (35.4 per cent) and agribulks (22.8 per cent). While manufactures and agribulks each increased by 6.0 per cent in 2014, metals and minerals declined by 3.0 per cent (*Dry Bulk Trade Outlook*, 2015a). Growth in manufactures reflected the firm increase in Chinese steel production and export growth supported by a tax rebate on some products as well as weaker domestic steel demand. Exports of metals and minerals were constrained by reduced Indonesian exports of nickel ore following the implementation of the export ban in January 2014. China's nickel ore imports are increasingly sourced from the Philippines, which have come to dominate the international nickel ore market in the past year. The drop in metals and minerals is also reflective of the fall in anthracite shipments resulting from a drop in Viet Nam's exports (Clarksons Research, 2015a).

Other dry cargo: Containerized trade

In 2014, global containerized trade was estimated to have increased by 5.3 per cent and reached 171 million TEUs (see figure 1.6 (a)). Global growth was boosted by the recovery on the headhaul journeys (peak legs) of the major East–West trans-Pacific and Asia–Europe trade lanes. Partly reflecting the recovery in the United States and the improved prospects for Europe, containerized trade volumes carried on the Asia–Europe and trans-Pacific peak legs are estimated to have increased by 7.5 per cent and 6.3 per cent respectively (Clarksons Research, 2015e). In comparison, and reflecting a weaker import demand in Asia, trade volumes on backhaul journeys remained weak. Weaker demand for imports from Europe and North America does not necessarily reflect a drop in the overall import demand, as imports into Asia often include waste and other residual products. Volumes on the westbound leg of the trans-Pacific route contracted while shipments on the eastbound leg of the Asia–Europe trade route increased only marginally (see table 1.7 and figure 1.6 (b)).

The recovery on the mainlane East–West routes does not, however, reveal the changing patterns of global demand. The total mainlane container trade is estimated to have grown by 9.0 per cent between 2007 and 2014 while trade volumes on the non-mainlane trades are said to have expanded by 45 per cent during the same period. Consequently, the share of world trade held by the mainlane trades fell from

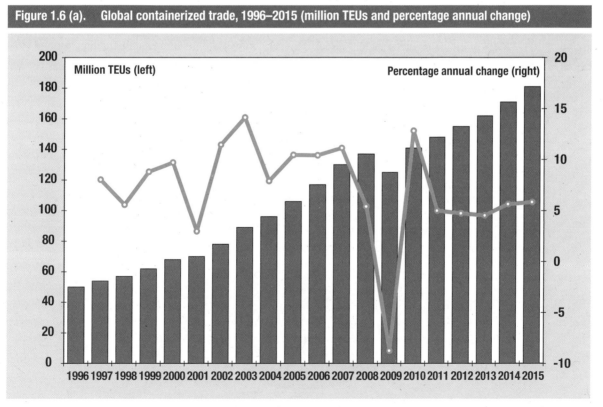

Figure 1.6 (a). Global containerized trade, 1996–2015 (million TEUs and percentage annual change)

Source: UNCTAD secretariat, based on Drewry Shipping Consultants, Container Market Review and Forecast 2008/2009; and Clarksons Research, *Container Intelligence Monthly*, various issues.

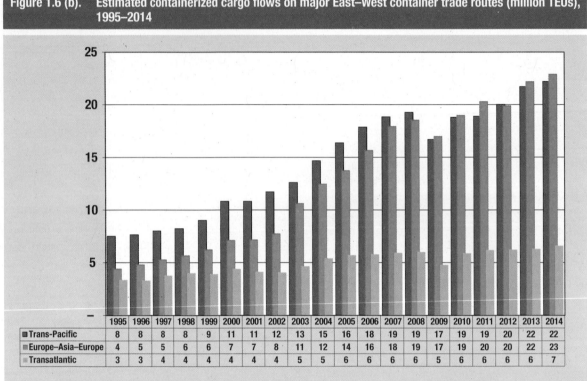

Figure 1.6 (b). Estimated containerized cargo flows on major East–West container trade routes (million TEUs), 1995–2014

	1995	1996	1997	1998	1999	2000	2001	2002	2003	2004	2005	2006	2007	2008	2009	2010	2011	2012	2013	2014
Trans-Pacific	8	8	8	8	9	11	11	12	13	15	16	18	19	19	17	19	19	20	22	22
Europe–Asia–Europe	4	5	5	6	6	7	7	8	11	12	14	16	18	19	17	19	20	20	22	23
Transatlantic	3	3	4	4	4	4	4	4	5	5	6	6	6	6	5	6	6	6	6	7

Source: UNCTAD secretariat, based on the Global Insight Database as published in *Bulletin FAL*, 288(8/2010) (International maritime transport in Latin America and the Caribbean in 2009 and projections for 2010), United Nations Economic Commission for Latin America and the Caribbean (ECLAC). Data for 2009, 2010, 2011, 2012, 2013 and 2014 are based on table 1.7.

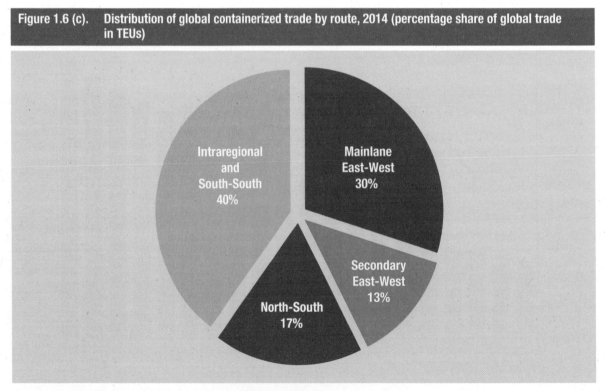

Figure 1.6 (c). Distribution of global containerized trade by route, 2014 (percentage share of global trade in TEUs)

- Intraregional and South-South 40%
- Mainlane East-West 30%
- North-South 17%
- Secondary East-West 13%

Source: UNCTAD secretariat, based on Clarksons Research (2015e); and *Lloyd's List Data Hub Statistics*, various issues.

CHAPTER 1: DEVELOPMENTS IN INTERNATIONAL SEABORNE TRADE

Table 1.7. Estimated containerized cargo flows on major East–West container trade routes, 2009–2014 (million TEUs and percentage annual change)

	Transpacific		Europe Asia		Transatlantic	
	Asia–North America	North America–Asia	Asia–Europe	Europe–Asia	Europe–North America	North America–Europe
2009	10.6	6.1	11.5	5.5	2.8	2.5
2010	12.3	6.5	13.3	5.7	3.2	2.7
2011	12.4	6.6	14.1	6.2	3.4	2.8
2012	13.1	6.9	13.7	6.3	3.6	2.7
2013	13.8	7.9	14.3	6.9	3.6	2.7
2014	14.7	7.5	15.4	7.0	3.9	2.7
Percentage change 2013–2014	6.3	-4.5	7.5	1.3	8.3	0.0

Source: UNCTAD secretariat, based on data from MDS Transmodal as published in Lloyd's List Data Hub Trade Statistics; and Containerisation International, various issues. Data for 2013 and 2014 are sourced from Clarksons Research, Container Intelligence Monthly, 17(4), May 2015.

36.0 per cent in 2007 to 30.0 per cent in 2014. At the same time, intraregional (led by intra-Asian trade) and South–South trade accounted for 40 per cent of global containerized volumes in 2014, followed by flows on the mainlane East–West (30 per cent), North–South (17 per cent) and secondary East–West trade routes (13 per cent) (Clarksons Research, 2015f) (figure 1.6 (c)).

Other relevant developments affecting containerized trade in 2014 included continued overcapacity, the cascading effect (ship capacity moved from main/ artery lanes to secondary routes), the uncertainty about the future of slow steaming (see also section B.1) and the alignment of major container ship operators in four mega-alliances.

The oversupply of container ship capacity remained a challenge given, in particular, the current cascading effect and related implications for port infrastructure requirements, the configuration of shipping services (direct versus trans-shipments), and earnings and profitability on the routes where ships were redeployed. There were also concerns about the continued dominance of very large vessels in the container ship order book and the mismatch between the delivery of high-capacity vessels and the pattern of global demand growth.

Initially implemented in response to higher oil and bunker fuel prices, slow steaming helped manage oversupply in container shipping. Slow steaming is estimated to have resulted in the employment of 1.3 million TEUs, the equivalent of 7.0 per cent of the global container fleet capacity (Ship & Bunker, 2014b). Despite the recovery on the main East–West trade lanes and the lower oil prices and bunker fuel costs, the practice of slow steaming in container shipping continued and appears to be the norm as there is no outright increase in vessel speeds (ShippingWatch, 2014). In the meantime, shipowners continue to order very large container ships, as illustrated by the very recent ordering of 11 second-generation Triple-E container vessels with a capacity of 19,630 TEUs each (Lloyd's List, 2015d).

Operators on the Far East to Europe trade route continued to pursue lower costs through vessel sharing arrangements and by deploying very large container ships. Four key alliances are now in operation and include 2M, the Ocean Three, the G6 and the CKYHE. The exact impact of this new alignment of the major container ship operators has yet to be fully assessed. Meanwhile, shippers are advocating greater scrutiny and the need to conduct reviews to determine how the alliances are impacting on the industry. In this respect, European shippers have launched an initiative to carry out a wide industry survey and a review of the implications of the mega-vessel sharing agreement (JOC staff, 2015).

C. SUSTAINABLE AND RESILIENT MARITIME TRANSPORT SYSTEMS

The year 2015 is a milestone for sustainable development. With the international community currently elaborating a post-2015 development agenda, there is a renewed opportunity to strengthen the international commitment to sustainable development and consider how best to mainstream sustainability principles across all economic sectors, including maritime transport.

With over 80 per cent of world merchandise trade being carried by sea, maritime transport remains the backbone of international trade and globalization. Equally, the sector is a key enabling factor for other sectors and economic activities such as marine equipment manufacturing, maritime auxiliary services (for example, insurance, banking, brokering, classification and consultancy), fisheries, tourism and the offshore energy sector, as well as other marine-based industries such as shipbuilding and ship demolition. In this context, sustainable maritime transport systems entail, among other factors, transport infrastructure and services that are safe, socially acceptable, universally accessible, reliable, affordable, fuel-efficient, environmentally friendly, low carbon and climate-resilient.

Achieving greater sustainability in transport, including maritime transport, has long been recognized as a key development objective, including in the context of the 1992 Earth Summit, the United Nations Conference on Sustainable Development, UNCTAD XIII, the third International Conference on Small Island Developing States (SIDS), the second United Nations Conference on Landlocked Developing Countries, and, more recently, the United Nations General Assembly resolution on the "Role of transport and transit corridors in ensuring international cooperation, stability and sustainable development" (A/RES/69/213). Additional momentum is also generated by the work carried out by the United Nations Secretary-General High-Level Advisory Group on Sustainable Transport. Established with a view to providing recommendations on sustainable transport that are actionable at global, national and local as well as sectoral levels, the High-Level Advisory Group is expected to publish a report on the global transport outlook and convene the first international conference on sustainable development in 2016.

Against this background the following sections highlight selected relevant issues that lie at the interface of maritime transport and sustainable development.

1. Factors driving sustainability in maritime transport

Efforts to improve the energy-related, environmental and social performance of the maritime transport sector are largely driven by regulation, including in particular rules adopted under the auspices of IMO. Sustainability and resilience-motivated regulations span a broad range of issues and include safety (accidents), security (regulatory measures and piracy), marine pollution (for example, oil spills, ballast water, garbage and ship paint), labour conditions (seafarers' rights and working conditions), air pollution (SOx) and nitrogen oxides (NOx)), as well as GHG emissions.

Market requirements and growing customer demands for greater corporate social responsibility in global supply chains, transparency, agility, reliability and lighter environmental footprints are also increasingly driving significant changes in the maritime transport industry. Customers across supply chains are

Box 1.1. Examples of voluntary self-regulation in shipping

- The Clean Cargo Working Group has developed tools and methodologies to help understand and manage sustainability impacts. Relevant measures include average trade lane emissions data that can be used for a benchmarking of carriers' performances based on their carbon emissions, as well as for more informed decisions by both carriers and shippers (Business for Social Responsibility, 2014).
- The World Ports Climate Initiative, under the International Association of Ports and Harbors (IAPH): The 50 participating ports in the Initiative are engaged in reducing GHG emissions from their activities, including by influencing the sustainability of supply chains. For example, the Environmental Ship Index aims to identify seagoing ships that have better performance in terms of reducing air emissions, and includes a reporting scheme on GHG emissions from ships. The Environmental Ship Index can be used to promote clean ships (IAPH, 2015a).
- IAPH Air Quality and Greenhouse Gas Tool Box, and work relating to climate adaptation in ports such as the Climate Protection Plan Development (IAPH, 2015b).
- The Sustainable Shipping Initiative brings together leading companies from across the industry and around the world with a view to a sustainable future. Relevant activities include the launch of the Case for Action report in 2011, and efforts to promote greater uptake of sustainable shipping rating schemes to provide transparency and comparability and to enable cargo owners, charters and shipowners to integrate sustainability into commercial decisions (Sustainable Shipping Initiative, 2015).
- Charterers representing 20 per cent of global shipped tonnage are adopting policies to avoid using inefficient ships based on their GHG emissions performance (*International Transport Journal*, 2015).

increasingly expecting transportation service providers, including maritime transport service providers, to act as strategic partners that can help them achieve economic benefits as well as value for the environment and society (Business for Social Responsibility, 2010).

In response to the growing demands both at the regulatory and the market levels the maritime transport industry is increasingly, in addition to regulation and mandated measures, taking voluntary measures and adopting private self-regulation to integrate sustainability and resilience principles into activities, policies and decisions. Box 1.1 illustrates some examples of actions taken at the industry level both in response to as well as in anticipation of greater demands for improved performances in terms of sustainability and resilience.

2. Access, connectivity and infrastructure

The strategic importance of maritime transport infrastructure and services for market access, globalized production, trade competitiveness, employment, income generation, poverty reduction and social progress cannot be overemphasized. Consequently, for many developing countries, addressing the physical and non-physical barriers such as infrastructure issues (for example, insufficiency, inadequacy, congestion and maintenance requirements), missing links and interoperability of, for example equipment, vehicles, technologies and standards, is key.

However, the transport infrastructure gap remains a significant challenge in many developing regions. Global transport infrastructure needs have been estimated at $11 trillion over the period 2009–2030 (OECD, 2011). Meanwhile, the infrastructure gap in developing countries, including transport infrastructure, is significant. In the Latin America and the Caribbean region for example, investment needs required annually to meet infrastructure demand for the period 2012–2020 are estimated at 6.2 per cent of GDP, or some $320 billion (ECLAC, 2014).

To close the gap on the large infrastructure deficit in developing countries, including in transportation, existing estimates indicate that spending must reach $1.8 trillion–$2.3 trillion per year by 2020 compared with the current levels of $0.8 trillion–$0.9 trillion per year (United Nations Development Programme (UNDP), 2013). Currently 60 per cent of estimated total annual transport infrastructure investments are allocated to OECD countries (Partnership on Sustainable Low Carbon Transport, 2015).

A well-articulated transport infrastructure vision and a long-term plan that also seeks to close the infrastructure gap in maritime transport should be pursued as matters of priority. Such efforts should be based on a careful coordination of the social, economic and physical development of maritime transport systems. Maritime transport infrastructure developers, investors and managers should mainstream sustainability and resilience criteria into their broader transport development plans at the early stages of the relevant decision-making and investment processes. As maritime transport infrastructure such as ports have long life cycles, not accounting for the long-term sustainability and resilience including climate resilience requirements, may involve costly retrofitting of equipment and infrastructure and adjustment of operations and services.

3. Energy and transport costs

As discussed in section B.1, the heavy reliance of maritime transport on fossil fuels for propulsion enhances the exposure of freight rates and transport costs to high oil price volatility. Although the mid-2014 drop in oil and bunker fuel prices may be a welcome development, the effect is likely to be short-lived, given the projected growth in the global energy demand and the risk of rapid cuts in oil production due to reduced investment in the oil extractive and refining industries.

An assessment of the effect of oil prices on maritime freight rates, including for containerized goods, iron ore and crude oil, reveals that rates and therefore transport costs in all three market segments were sensitive to a rise in oil prices, albeit at different degrees (UNCTAD, 2010). For containerized trade, the estimated elasticity ranges between 0.19 and 0.36; a similar elasticity is estimated for crude oil cargo – 0.28. For iron ore, on the other hand, the elasticity is estimated to be much larger, approximately equal to unity. Developing countries are already facing disproportionately higher transport costs, with UNCTAD estimating the 2013 average freight costs as a share of imports value at close to 7.0 per cent for developed economies, 10.0 per cent for developing economies and 8.0 per cent for the world average. The negative implications of volatile oil and fuel costs for economies' sustainable development can be significant given the potential impact on transport costs, affordability of services and trade competitiveness. Greater sustainability in maritime transport requires as a matter of priority that the overdependence on oil-based propulsion systems be effectively addressed (UNCTAD, 2010). Reducing exposure to volatility in oil prices and fuel costs though investment in energy efficiency measures, alternative energy sources and more sustainable operational and management practices can help control fuel and transport costs, derive efficiency gains and therefore enable more effective access to markets and promote trade competiveness.

4. Energy, environment and carbon emissions

In addition to raising transport costs and acting as a barrier to trade, heavy reliance on oil for propulsion undermines resource conservation objectives and leads to environmental deterioration through air and marine pollution and carbon emissions. In 2012, carbon dioxide (CO_2) emissions from international shipping were estimated at 2.2 per cent of global CO_2 emissions (IMO, 2014a). While the contribution of international shipping to global carbon emissions may be relatively low when assessed per unit of cargo and distance travelled, these emissions are, however, likely to grow if left unchecked. Forecast scenarios for the medium term suggest that international shipping carbon emissions could increase 50–250 per cent by 2050, depending on economic growth and global energy demand. Equally, international freight, including maritime transport, is projected to more than quadruple by 2050, with associated CO_2 emissions generated by all modes engaged in the international trade between 2010 and 2050 growing by a factor of 3.9 (International Transport Forum/OECD, 2015). In this context, locking in fossil fuels and related technologies in freight transport, including maritime transport, will perpetuate unsustainable transport patterns.

Breaking away from fossil fuel-intensive maritime transport systems and shifting towards greater sustainability and resilience, including through tailored and targeted policies, regulations, incentives and programmes, is an imperative for freight transport, including maritime. Relevant strategies for the freight transport sector include, for example: promoting, when feasible and as applicable, a modal shift towards more environmentally and less energy-intensive modes (maritime, short-sea shipping, waterways and rail); shifting to lower-carbon fuels; promoting infrastructure maintenance and management; rethinking supply-chain designs, including the location of production sites; reshaping transport architecture and networks and rerouting trade to ensure the most energy-efficient and less carbon-emitting trajectories; improving cooperation and stakeholder networking; promoting trade facilitation measures that reduce border delays and inefficiencies; making greater use of information and communications technologies as well as intelligent transport systems; and promoting energy-efficient transport technologies.

The potential benefits of energy efficiency measures can be significant. The International Energy Agency considers energy efficiency as the world's "first fuel" and estimates the 2012 global investment markets in energy efficiency at between $310 billion and $360 billion (Kojima and Ryan, 2010). A significant potential for energy efficiency exists in emerging economies outside the OECD, with efficiency able to slash up to $90 billion in global transport-related fuel costs by 2020 while reducing local air pollution. In maritime transport, key regulatory instruments addressing the nexus between energy, air pollution and carbon emissions from shipping are the technical and operational measures mandated by IMO in 2011 (IMO, 2015). The relevant requirements include the Energy Efficiency Design Index (EEDI) and the Ship Energy Efficiency Management Plan (SEEMP). In considering 22 potential ship efficiency measures and calculating their aggregated cost effectiveness and reduction potential, one study finds that, by 2020, the industry's growing fleet could reduce annual CO_2 emissions by 33 per cent of the projected annual total (International Council for Clean Transportation, 2011). Another study investigated 28 energy-saving options and estimated a reduction of CO_2 emissions in shipping by 2030 of more than 50 per cent (Alvik et al., 2010). Other relevant measures include those relating to the IMO sulphur limits imposed on fuels used by ships, globally as well as in designated ECAs (see section B).

5. Climate change impacts, adaptation and resilience-building

Maritime transport is facing the dual challenge of climate mitigation and adaptation.[1] While future trends in emissions from international shipping remain uncertain (subject to international efforts/commitments to cut GHG emissions and the efforts of IMO and the twenty-first session of the Conference of the Parties to the United Nations Framework Convention on Climate Change (COP21) and curbing GHG emissions remains an urgent imperative to ensure manageable global warming levels, the effects of climate variability and change – irrespective of the causes – are already being felt in different parts of the world, often in the poorest countries with low adaptive capacity.

Transport networks and seaports in particular are likely to be highly affected by climate change factors given port location and vulnerability. Climatic factors such as rising water levels, floods, storms, precipitation, extreme weather events and associated risks such as coastal erosion, inundation and deterioration of hinterland connections have implications for shipping volumes and costs, cargo loading and capacity, sailing and/or loading schedules, storage and warehousing. With international trade being increasingly multimodal and requiring the use of rail, road and waterway transport, these impacts will also affect the transport corridors above and beyond the ports acting as gateways.

Climate change impacts on maritime transport can be direct and indirect, that is, by causing changes in demand for maritime transport services (Gledhill et al., 2013). In this respect, one study estimated that in 2005, the exposure of 136 port megacities to coastal flooding (population and assets) was $3 trillion (Nicholls et al., 2008). When assuming a sea level rise of half a metre by 2050 (the tipping scenario), the asset exposure (that is, economic assets in the form of buildings, transport infrastructure, utility infrastructure, physical assets within built infrastructure, vehicles and other assets) of the same 136 port megacities was projected to be $28 trillion (Lenton et al., 2009). A climate-induced port closure or disruption to operations can be costly, although, to put things in perspective, a comparison with, for example, the impact of a labour dispute-related port closure can be made.

Building the climate resilience of maritime transport systems is therefore a precondition for sustainability. Enhancing understanding and technical knowledge among policymakers, transport planners and transport infrastructure managers of climate change impacts on coastal transport infrastructure, services and operations is of the essence. It is equally important to strengthen their capacity to take informed decisions and respond with effective, appropriate and well-designed climate policy and adaptation measures. Conducting risk assessments for critical transport infrastructure and facilities, especially in ports, will be crucial to ensure that any adaptation measures adopted are tailored to reflect the local conditions, especially in developing regions. However, to be more effective, enhancing adaptive capacity requires that actions are also integrated with other policies such as disaster preparedness, land-use planning, environmental conservation, coastal planning, and national plans for sustainable development.

6. Financing sustainable and resilient maritime transport

Enhancing the sustainability and resilience of maritime transport entails some cost implications and calls for additional resources. However, in an era of increasingly constrained national budgets, finding innovative ways to mobilize the requisite sources is critical. New sources and mechanisms and greater private sector involvement such as through public–private partnerships is important. In terms of innovative financing mechanisms, climate finance could emerge as an important channel for mobilizing additional resources, including for maritime transport. In this respect, at their June 2015 summit, the Group of Seven leaders reiterated their commitment to the Copenhagen Accord to jointly mobilize $100 billion per year by 2020, and to make the green climate fund operational in 2015 (Group of Seven Summit, 2015). Some analysts argue that, in connection with climate action, redirecting existing resources towards low-carbon and sustainable uses would be sufficient (Vivid Economics, 2014). The argument is as follows: the best estimate of additional investments required to mitigate and adapt to climate change in developing regions amounts to $400 billion–$500 billion per year by 2030. At the same time, overall investment in the same countries increased by more than $3.25 trillion during the period 2002–2012. Therefore, redirecting just a fraction of the expected continued growth in investment towards mitigation and adaptation action would support the realization of climate and sustainable development objectives (Vivid Economics, 2014).

In addition to increasing the levels and diversifying the sources of finance, financing energy-efficient maritime transport systems requires that the key barriers to investments such as the split incentives involving shipowners and charterers (charters do not share or give back savings to shipowners) be addressed. With investment in ship energy efficiency being generally carried out by shipowners/operators, the costs associated with leveraging innovative ship energy-efficient technologies and alternative fuels (for example, equipment, hull design, engines, propulsion systems and operational measures) are part of the overall capital costs involved in ordering a ship. Decisions to invest, for example, in eco-ships that save on fuel use and reduce air emissions but that are more expensive, are made by shipowners/operators who depend largely on the banking sector to meet their financing requirements. On the positive side, banks are said to be increasingly taking into account sustainability criteria and ship energy-efficiency performances, in particular when making financing decisions. With energy-efficient ships being more likely to have a higher asset value and a longer lifespan, banks are reported to be increasingly favouring investments in sustainable ships such as eco-ships that entail reduced financing risks (including better chartering potential and lower fuel costs) (*The Marine Professional*, 2015).

Shipping-related market-based instruments could also be used to help finance investments in energy efficiency. At present and in addition to technical design standards, the international community, under the auspices of IMO/United Nations Framework Convention on Climate Change (UNFCCC), is considering several instruments to regulate GHG emissions from international shipping, including market-based measures such as levies/taxes and emission-trading mechanisms. Revenues generated by these instruments could be reinvested in the shipping sector, including with a view to energy efficiency measures. However, so far agreement on any international market-based instrument to regulate carbon emissions from international shipping has yet to be achieved.

Governments have a role to play in supporting private sector investment in energy-efficient technologies and alternative fuels by creating a favourable climate, including through fiscal and monetary incentives (for example, tax breaks and subsidies in support of energy-efficient technologies, grants or subsidies for research and development) and enabling regulatory and policy frameworks that support innovation and facilitate processes and procedures. Governments can also, in cooperation, for example, with the shipping and port industry, partner to leverage carbon markets to promote energy-efficient technologies. As has been argued in the case of air transport, development banks also have a role to play (World Bank/International Bank for Reconstruction and Development, 2012). For example, they could support energy efficiency measures that apply to maritime transport infrastructure (for example, technologies that support cold ironing in ports) to complement ship energy-efficiency measures.

To sum up, the year 2015 is a milestone for sustainable development, in which a path for a new international sustainable development agenda will be set, and a global climate policy framework adopted. Maritime transport has an important role to play in addressing the global sustainability and resilience agenda. The sector is thus at a critical juncture as it has an opportunity to assert its strategic importance as an economic activity that generates employment and revenue, enables trade, supports supply chains and links communities; and underscore its potential to generate value in terms of economic viability as well as social equity, resource conservation and environmental protection. However, for this role to effectively materialize, relevant sustainability and resilience criteria need to be integrated and mainstreamed into maritime transport planning, policies and investment decisions. Adopting a multi-stakeholder approach involving Governments, the maritime transport industry, financial institutions and other relevant partners is an overriding imperative for these efforts to be successful. Equally, collecting, sharing and disseminating relevant data, including relevant sustainability and performance indicators, is necessary, as is the need to scale up financing, enhance capacity-building, share best practices and enable greater use of relevant technologies.

Chapter 2 addresses trends in the world merchant fleet, chapter 4 deals with port-related developments and chapter 5 considers legal issues and regulatory developments, each highlighting ways in which the maritime transport industry can contribute to achieving greater sustainability in the maritime transport sector.

REFERENCES

Alvik S, Eide M, Endersen Ø, Hoffmann P and Longva T (2010). Pathways to low carbon shipping. Abatement potential towards 2030. Det Norske Veritas. February.

Barnard B (2015). Low oil prices, shipper pushback nullify low sulfur's impact. *Journal of Commerce*. 16 February.

British Petroleum (2015). *Statistical Review of World Energy 2015*.

Business for Social Responsibility (2010). *Supply Chain Sustainability: A Practical Guide for Continuous Improvement*. United Nations Global Compact and Business for Social Responsibility.

Business for Social Responsibility (2014). Global maritime trade lane emissions factors. Available at http://www.bsr.org/reports/BSR_CCWG_Trade_Lane_Emissions_Factors.pdf (accessed 9 September 2015).

Clarksons Research (2015a). *Shipping Review and Outlook*. Spring.

Clarksons Research (2015b). *Seaborne Trade Monitor*. 2(6). June.

Clarksons Research (2015c). *Oil and Tanker Trade Outlook*. May.

Clarksons Research (2015d). *Oil and Tanker Trade Outlook*. January.

Clarksons Research (2015e). *Container Intelligence Monthly*. 17(6). June.

Clarksons Research (2015f). *Container Intelligence Quarterly*. First quarter.

Cohen MA and Lee HL (2015). Global supply chain benchmark study: An analysis of sourcing and re-structuring decisions. Supply Chain Navigator. April. Available at http://scnavigator.avnet.com/article/april-2015/global-supply-chain-benchmark-study/ (accessed 9 September 2015).

Dry Bulk Trade Outlook (2015a). Clarksons Research. May.

Dry Bulk Trade Outlook (2015b). Clarksons Research. January.

ECLAC (2014). Investment in infrastructure in Latin America and the Caribbean. Available at http://www.cepal.org/sites/default/files/infographic/files/infraestructura_espanol.pdf (in Spanish) (accessed 9 September 2015).

Elliott-Green N (2015). Crude trade: Looking beyond the barrels. Clarksons Research. January.

European Commission (2001). Directive 2001/80/EC of the European Parliament and of the Council on the limitation of emissions of certain pollutants into the air from large combustion plants. October.

Francois J, Manchin M, Norberg H, Pindyuk O and Tomberger P (2013). Reducing transatlantic barriers to trade and investment: An economic assessment. Centre for Economic Policy Research. London.

Gledhill R, Hamza-Goodacre D and Ping L (2013). Business-not-as-usual: Tackling the impact of climate change on supply chain risk. PricewaterhouseCoopers.

Group of Seven Summit (2015). Think Ahead, Act Together. Group of Seven Summit Declaration. June. Available at https://www.g7germany.de/Content/EN/Artikel/2015/06_en/g7-gipfel-dokumente_en.html (accessed 9 September 2015).

HSBC Bank (2015). Global connections – Global overview. Trade forecast reports.

IAPH (2015a). World Ports Climate Initiative. June. Available at http://wpci.iaphworldports.org/ (accessed 7 September 2015).

IAPH (2015b). IAPH Tool Box for Port Clean Air Program. Available at wpci.iaphworldports.org/iaphtoolbox/ (accessed 9 September 2015).

International Council for Clean Transportation (2011). Reducing greenhouse gas emissions from ships: Cost effectiveness of available options. White paper. Available at http://www.theicct.org/sites/default/files/publications/ICCT_GHGfromships_jun2011.pdf (accessed 9 September 2015).

International Energy Agency (2015). Oil market report. June.

IMO (2014). Third IMO GHG Study 2014 – Final report. MEPC 67/INF.3. London.

IMO (2015). Prevention of air pollution from ships. Available at http://www.imo.org/en/OurWork/Environment/PollutionPrevention/AirPollution/Pages/Air-Pollution.aspx (accessed 9 September 2015).

International Monetary Fund (2015). Learning to live with cheaper oil amid weaker demand. Regional Economic Outlook Update. Washington, D.C.

International Transport Forum/OECD (2015). *ITF Transport Outlook 2015*. Paris.

International Transport Journal (2015). Charterers to exclude inefficient vessels. 29 May.

JOC (2014). Falling bunker price gets industry talking about speeding up ships. 3 November.

JOC staff (2015). European shippers launch global review of mega-alliances. *JOC*. 23 April.

Johnson S (2015). Oil price drop wreaks havoc on Russian economy. *Market Realist*. 30 January.

Jones D and Worthington B (2014). Europe's failure to tackle coal risks for the EU low-carbon transition. Sandbag Climate Campaign.

Kojima K and Ryan L (2010). Transport energy efficiency. Implementation of IEA recommendation since 2009 and next steps. September. International Energy Agency. Available at https://www.iea.org/publications/freepublications/publication/transport_energy_efficiency.pdf (accessed 9 September 2015).

Lenton T, Footitt A and Dlugolecki A (2009). Major tipping points in the Earth's climate system and consequences for the insurance sector. World Wide Fund for Nature, Gland, and Allianz SE, Munich.

Lloyd's List (2015a). Shippers' calls for faster Asia–Europe services fall on deaf ears. 21 April.

Lloyd's List (2015b). Carriers warned of impact of speeding up services. 10 February.

Lloyd's List (2015c). LNG volumes forecast to pick up substantially in 2015 and 2016. 10 April.

Lloyd's List (2015d). Maersk Line orders 11 ultra-large container vessels. 3 June.

Nicholls RJ, Hanson S, Herweijer C, Patmore N, Hallegatte S, Corfee-Morlot J, Château J and Muir-Wood R (2008). Ranking port cities with high exposure and vulnerability to climate extremes exposure estimates. Environment Working Papers No. 1. OECD.

OECD (2011). *Strategic Transport Infrastructure Needs to 2030*. Paris.

Partnership on Sustainable Low Carbon Transport (2015). Transport at COP20: Despite limited leaps, Lima limps. Climate finance as the engine for more low-carbon transport. Partnership on Sustainable Low Carbon Transport and Bridging the Gap Initiative.

Petri PA and Plummer MG (2012). The trans-Pacific partnership and Asia–Pacific integration: Policy implications. Policy brief No. PB12-6. Peterson Institute for International Economics. Washington, D.C.

Politico Magazine (2014). What the 2014 oil crash means. Prices are falling – fast. Is that good or bad news for the United States? 16 October.

Ship & Bunker (2014a). Falling oil prices push owners to offload ECO ships. 15 December.

Ship & Bunker (2014b). Alphaliner: Slow steaming keeps 7% of global fleet employed. 24 October.

ShippingWatch (2014). Maersk Line sticks to slow steaming. 22 October.

Sustainable Shipping Initiative (2015). Available at http://ssi2040.org (accessed 9 September 2015).

The Marine Professional (2015). Banks more likely to finance efficient ships. 22 April.

Trimmel B (2015). Iron ore exports: A dangerous race? Shipping Intelligence Network. April.

UNCTAD (2010). Oil prices and maritime freight rates: An empirical investigation. UNCTAD/DTL/TLB/2009/2. 1 April.

UNDP (2013). *Human Development Report 2013. The Rise of the South: Human Progress in a Diverse World*. New York.

Vivid Economics (2014). Financing green growth. Available at http://www.vivideconomics.com/publications/financing-green-growth (accessed 9 September 2015).

World Bank/International Bank for Reconstruction and Development (2012). Air transport and energy efficiency. Transport papers No. TP-38. Washington, D.C.

World Steel Association (2015). World crude steel output increases by 1.2% in 2014. January.

ENDNOTES

[1] For additional information about the science of climate change and its impacts on transport, including coastal transport infrastructure, see relevant documentation on UNCTAD work in the field available at http://unctad.org/en/Pages/DTL/TTL/Legal/Climate-Change-and-Maritime-Transport.aspx.

2

STRUCTURE, OWNERSHIP AND REGISTRATION OF THE WORLD FLEET

The world fleet grew by 3.5 per cent during the 12 months to 1 January 2015, the lowest annual growth rate in over a decade. In total, at the beginning of the year, the world's commercial fleet consisted of 89,464 vessels, with a total tonnage of 1.75 billion dwt. For the first time since the peak of the shipbuilding cycle, the average age of the world fleet increased slightly during 2014. Given the delivery of fewer newbuildings, combined with reduced scrapping activity, newer tonnage no longer compensated for the natural aging of the fleet.

Greece continues to be the largest ship-owning country, followed by Japan, China, Germany and Singapore. Together, the top five ship-owning countries control more than half of the world tonnage. Five of the top 10 ship-owning countries are from Asia, four are European and one is from the Americas.

The Review of Maritime Transport further illustrates the process of concentration in liner shipping. While the container-carrying capacity per provider per country tripled between 2004 and 2015, the average number of companies that provide services from/to each country's ports decreased by 29 per cent. Both trends illustrate two sides of the same coin: as ships get bigger and companies aim at achieving economies of scale, there remain fewer companies in individual markets.

New regulations require the shipping industry to invest in environmental technologies, covering issues such as emissions, waste, and ballast water treatment. Some of the investments are not only beneficial for the environment, but may also lead to longer-term cost savings, for example due to increased fuel efficiency.

Economic and regulatory incentives will continue to encourage individual owners to invest in modernizing their fleets. Unless older tonnage is demolished, this would lead to further global overcapacity, continuing the downward pressure on freight and charter rates. The interplay between more stringent environmental regulations and low freight and charter rates should encourage the further demolition of older vessels; this will not only help reduce the oversupply in the market, but also contribute to lowering the global environmental impact of shipping.

A. STRUCTURE OF THE WORLD FLEET

1. World fleet growth and principal vessel types

Responding to the growth in demand (see chapter 1), the world fleet grew by 3.5 per cent during the 12 months to 1 January 2015, the lowest annual growth rate in over a decade.[1] In total, at the beginning of the year, the world's commercial fleet consisted of 89,464 vessels, with a total tonnage of 1.75 billion dwt (figure 2.1 and table 2.1). The new tonnage added to the global fleet continued to decline in comparison with previous years in absolute terms. At the same time, the overall growth rate of tonnage was still above that of global GDP and trade growth, and even slightly higher than that of the growth of seaborne trade.

The greatest and expanding share in the global fleet are dry bulk carriers, which by the beginning of 2015 had reached a share of 43.5 per cent of total capacity; the result of a 4.4 per cent growth rate between 2014 and 2015 and even higher expansion in the years 2010–2013 (figure 2.2).

Despite the continued economic crisis, the container ship fleet increased by 5.2 per cent in the same period and thus stands in contrast to the slowdown in global economic growth. A further increase in the rate of containerization may to some extent lead to growing demand for container-carrying capacity, yet overall during recent years demand has grown less than supply, leading to a situation of continued oversupply in the container shipping market, resulting in continued downward pressure on container freight rates (see chapter 3).

The growth in the offshore and gas tanker segment surpassed all other vessel types and reflects the expansion of trade in gas and new offshore exploration projects. This development contrasts the slow growth in oil carriers (1.4 per cent). The ferries and passenger vessels fleet expanded by 4.8 per cent, indicating positive expectations about demand in the cruise industry. The overall positive development in the market segment of other types also indicates the further specialization of the global fleet (table 2.1).

The cyclical nature of shipbuilding is exemplified in figure 2.3, which illustrates the year the vessels built in 2014 were contracted. As illustrated in figure 2.4, total

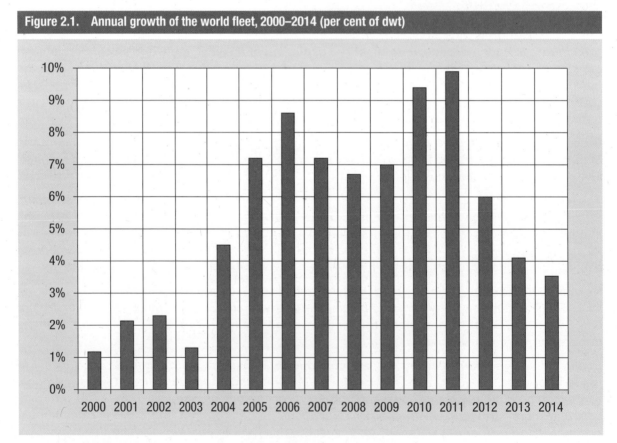

Figure 2.1. Annual growth of the world fleet, 2000–2014 (per cent of dwt)

Source: UNCTAD, *Review of Maritime Transport*, various issues.

CHAPTER 2: STRUCTURE, OWNERSHIP AND REGISTRATION OF THE WORLD FLEET

tonnage delivered in 2014 was only slightly more than half the tonnage delivered in 2011, the peak year of the historically largest ever shipbuilding cycle. Several years pass between the placement of an order for a new ship and the moment the new ship is delivered to the market. Ships are often ordered when the market is perceived as strong, only to be delivered years later, when the market may have become weaker.

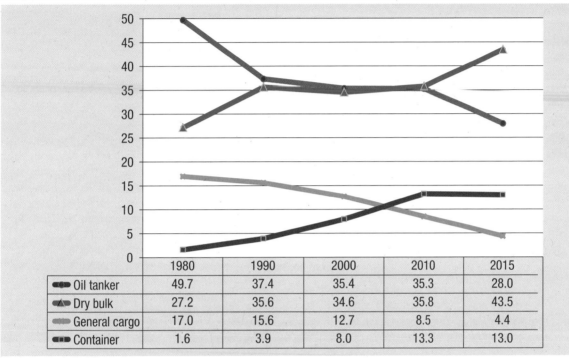

Figure 2.2. World fleet by principal vessel types, 1980–2015 (beginning-of-year figures, percentage share of dwt)

	1980	1990	2000	2010	2015
Oil tanker	49.7	37.4	35.4	35.3	28.0
Dry bulk	27.2	35.6	34.6	35.8	43.5
General cargo	17.0	15.6	12.7	8.5	4.4
Container	1.6	3.9	8.0	13.3	13.0

Source: UNCTAD secretariat, based on data supplied by Clarksons Research and the Review of Maritime Transport, various issues.
Note: All propelled seagoing merchant vessels of 100 GT and above, excluding inland waterway vessels, fishing vessels, military vessels, yachts and offshore fixed and mobile platforms and barges (with the exception of floating production, storage and offloading units (FPSOs) and drillships).

Table 2.1. World fleet by principal vessel types, 2014–2015 (beginning-of-year figures, thousands of dwt; percentage share in italics)

Principal types	2014	2015	Percentage change 2015/2014
Oil tankers	482 447	489 388	1.4%
	28.6%	28.0%	
Bulk carriers	728 322	760 468	4.4%
	43.1%	43.5%	
General cargo ships	77 507	76 731	-1.0%
	4.6%	4.4%	
Container ships	215 880	227 741	5.5%
	12.8%	13.0%	
Other types:	185 306	194 893	5.2%
	11.0%	11.1%	
Gas carriers	46 335	49 675	7.2%
	2.7%	2.8%	
Chemical tankers	41 688	42 181	1.2%
	2.5%	2.4%	
Offshore	69 513	74 174	6.7%
	4.1%	4.2%	
Ferries and passenger ships	5 531	5 797	4.8%
	0.3%	0.3%	
Other/n.a.	22 241	23 066	3.7%
	1.3%	1.3%	
World total	1 689 462	1 749 222	3.5%
	100%	100%	

Source: UNCTAD secretariat, based on data supplied by Clarksons Research.
Note: Propelled seagoing merchant vessels of 100 GT and above.

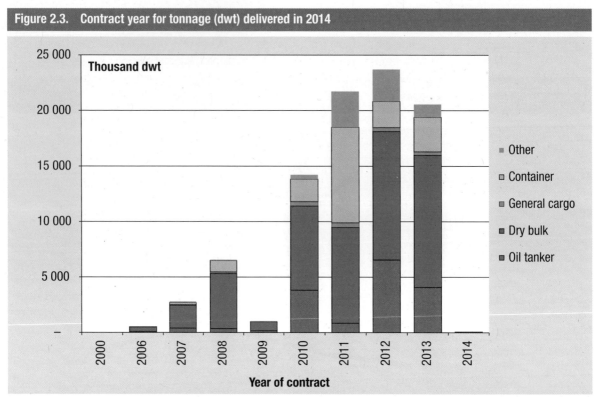

Figure 2.3. Contract year for tonnage (dwt) delivered in 2014

Source: UNCTAD secretariat, based on data supplied by Clarksons Research.
Note: Propelled seagoing vessels of 100 GT and above.

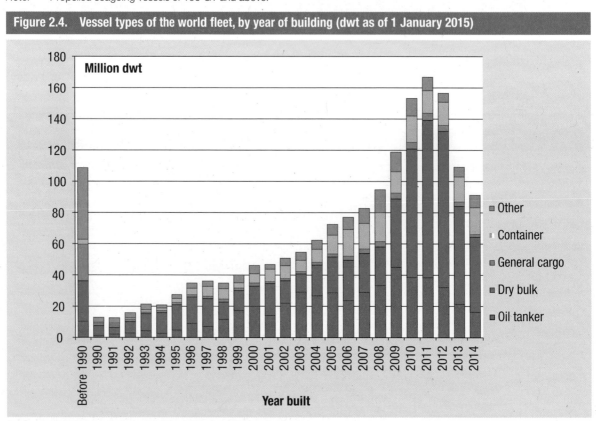

Figure 2.4. Vessel types of the world fleet, by year of building (dwt as of 1 January 2015)

Source: UNCTAD secretariat, based on data supplied by Clarksons Research.
Note: Propelled seagoing vessels of 100 GT and above.

CHAPTER 2: STRUCTURE, OWNERSHIP AND REGISTRATION OF THE WORLD FLEET

Table 2.2. Age distribution of the world merchant fleet, by vessel type, as of 1 January 2015 (percentage of total ships and dwt)

Country grouping Types of vessel		0–4 years	5–9 years	10–14 years	15–19 years	20 + years	Average age 2014	Average age 2015	Change 2015/2014
World: Bulk carriers	Ships	47.50	18.68	11.12	11.55	11.15	9.07	9.15	-0.09
	Dwt	51.88	18.73	10.46	9.94	8.99	8.08	7.98	0.10
	Average vessel size (dwt)	80 338	73 728	69 145	63 323	59 290			
World: Container ships	Ships	20.94	34.31	17.61	17.55	9.60	10.88	10.70	0.18
	Dwt	34.88	34.22	16.58	10.18	4.14	8.23	8.19	0.04
	Average vessel size (dwt)	74 310	44 487	42 001	25 869	19 235			
World: General cargo	Ships	10.68	14.89	7.70	8.96	57.76	24.86	24.18	0.68
	Dwt	22.09	18.86	10.05	10.17	38.83	17.97	17.76	0.21
	Average vessel size (dwt)	8 297	5 388	6 086	4 885	2 758			
World: Oil tankers	Ships	18.74	21.72	12.69	8.32	38.54	18.37	17.92	0.45
	Dwt	29.90	32.59	22.83	10.04	4.64	8.98	8.51	0.47
	Average vessel size (dwt)	83 196	78 871	95 231	65 702	6 521			
World: Others	Ships	16.55	16.87	9.22	8.88	48.48	22.22	21.86	0.36
	Dwt	20.41	26.49	12.31	9.16	31.62	15.65	15.30	0.35
	Average vessel size (dwt)	6 619	8 547	7 574	5 834	3 962			
World: All ships	Ships	14.94	15.64	8.35	7.96	53.12	20.25	19.89	0.35
	Dwt	38.71	25.50	14.90	9.92	10.97	9.63	9.41	0.22
	Average vessel size (dwt)	42 873	30 899	34 042	23 160	6 095			
Developing economies: All ships	Ships	20.28	17.71	8.64	9.24	44.12	19.76	19.43	0.33
	Dwt	41.55	20.45	10.97	10.98	16.05	10.37	10.20	0.17
	Average vessel size (dwt)	36 453	21 879	25 241	22 128	6 788			
Developed economies: All ships	Ships	20.20	21.02	12.79	11.24	34.76	18.52	18.17	0.35
	Dwt	37.46	29.00	17.56	9.10	6.88	8.90	8.65	0.25
	Average vessel size (dwt)	52 026	39 690	40 847	24 649	7 142			
Countries with economies in transition: All Ships	Ships	7.29	7.71	3.68	4.03	77.30	28.82	28.12	0.70
	Dwt	20.21	22.70	15.56	12.57	28.97	15.56	15.03	0.53
	Average vessel size (dwt)	17 659	20 706	27 366	20 029	2 398			

Source: UNCTAD secretariat, based on data supplied by Clarksons Research.
Note: Propelled seagoing vessels of 100 GT and above.

Most tonnage delivered in 2014 had been contracted during the previous four years, as well as to some extent in 2008 and 2007. Relatively fewer new orders were placed in 2009, after the economic slump (figure 2.3). Thus most current deliveries result from decisions made after the economic crisis. The continued high level of growth of container vessels indicates the industry's persistent strategy to realize economies of scale as well as cost savings, for example through increased energy efficiency.

The resulting oversupply of tonnage may not be good news for shipowners. However, it is a positive development from the perspective of those who aim at reviving global trade; there is no shortage of carrying capacity, and as a result trade costs continue to decline in the long term (see also chapter 3).

2. Age distribution of the world merchant fleet

For the first time since the peak of the shipbuilding cycle, the average age of the world fleet increased slightly during 2014. Given the delivery of fewer newbuildings, combined with reduced scrapping activity, newer tonnage no longer compensated for the natural aging of the fleet (table 2.2). As overall growth rates have been slowing down for the third consecutive year, the current aging of the fleet is a natural phenomenon of the concluding shipping cycle and will accelerate over the next few years. However, the current fleet is significantly younger than a decade ago. Average values somewhat hide that the low average fleet age is largely the result of newbuildings in the dry bulk and container sector, while the age of other vessel types continues to increase. The average age of "other" vessels is double that of the two previously mentioned sectors.

The distribution is also not equal across regions, countries and shipping routes. A central driver for these differences is the cascading effect induced by overcapacity in the main trade lanes, which shifts older and often smaller vessels to secondary routes. Also, new environmental regulations push older tonnage into regions with less restrictive regimes. Peripheral and less developed regions, and particularly services between these, already tend to be those with the oldest and potentially less environmentally friendly fleets. Hence, the cascading effect actually has a positive impact from an environmental perspective, as it pushes relatively more modern vessels into the peripheral regions and routes. As these ships tend to be bigger, this trend increases the pressure on port infrastructure development in developing countries.

3. Environmental sustainability: Trends in vessel technologies

New regulations (see also chapter 5) require that the shipping industry invest in environmental technologies, covering issues such as emissions, waste and/or ballast water treatment. Some of the investments are not only beneficial for the environment, but may lead to longer term cost savings, for example thanks to increased fuel efficiency.

Figure 2.5 illustrates the increasing introduction of ballast water treatment systems, making use of technologies such as ultraviolet, chemical and filtration systems. Their effectiveness varies according to factors such as seawater salinity, temperature and sediment load (Clarksons Research, 2014a). In 2013 and 2014, more than half of new container ships were built with such systems. The share was lower, albeit also growing, in other vessel types.

Emissions from maritime transport are of increasing concern. More stringent measures have been adopted by IMO in relation to SOx and NOx. As regards SOx, there exist new global limits, as well as more stringent limits in ECAs in Europe and North America.

As regards technologies, there are three main methods of compliance with the SOx regulations. These are (a) low sulphur fuels such as marine gas oil; (b) scrubber technology for the after-treatment of exhaust gas that uses seawater to wash out SOx; (c) alternative fuels, notably LNG, and potentially also biofuels and methanol.

The solution an owner chooses for will depend on a range of factors, including the amount of time spent in ECAs, the ship's fuel consumption and its age. Scrubber systems reportedly cost in the range of $2 million–$4 million per vessel (Clarksons Research, 2014b) and it is expected that the majority of shipowners will switch to marine gas oil in the short term. Only for ships operating mostly in ECAs will scrubbers become more economic, as they enable the use of standard heavy fuel oil, which is cheaper than low sulphur fuel alternatives.

Data on vessel newbuildings suggests that the majority of ships will meet new limits for ECAs by

CHAPTER 2: STRUCTURE, OWNERSHIP AND REGISTRATION OF THE WORLD FLEET

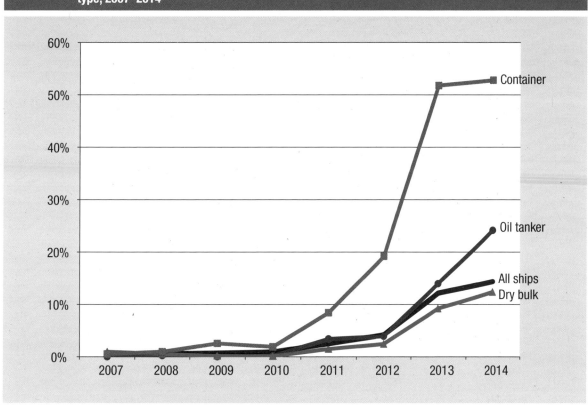

Figure 2.5. Share of newbuildings (number of ships) with ballast water treatment systems, by main vessel type, 2007–2014

Source: UNCTAD secretariat, based on data supplied by Clarksons Research.
Note: Propelled seagoing merchant vessels of 1,000 GT and above.

switching to low sulphur fuels such as marine gas oil in the short term. A small share of the existing fleet and the order book is reported to have scrubbers installed. In particular, ships that spend a lot of time in ECAs find the installation of scrubbers convenient. In the longer term, as global SOx limits are further tightened, it can be expected that more scrubbers will be installed rather than opting for the short-term solution of using marine gas oil (Clarksons Research, 2014b).

Using LNG as fuel is another option to reduce emissions. In March 2015, only 178 ships were LNG fuelled or capable of running on LNG, most of which were LNG carriers themselves (Clarksons Research, 2015a). Nevertheless, the share of tonnage using LNG as fuel is increasing, and as regulations on emissions become more stringent, this growth can be expected to continue in the longer term. The use of LNG as fuel will also depend on the installation of the corresponding bunkering infrastructure. Currently, the infrastructure is lacking, with only sparse coverage of LNG fuelling stations, mostly located in Northern Europe (Morgan Stanley, 2013).

B. OWNERSHIP AND OPERATION OF THE WORLD FLEET

1. Ship-owning countries

Greece continues to be the largest ship-owning country, accounting for more than 16 per cent of the world total, followed by Japan, China, Germany and Singapore. Together, the top five ship-owning countries control more than half of the world tonnage (dwt) (table 2.3). Five of the top ten ship-owning countries are from Asia, four are European, and one (the United States) is from the Americas.

Over the last decade, China, Hong Kong (China), the Republic of Korea and Singapore have moved up in the ranking of largest ship-owning countries, while Germany, Norway and the United States have a lower market share today than in 2005.

In South America, the largest ship-owning country (in dwt) continues to be Brazil, followed by Mexico, Chile and Argentina. The African country with the largest

Table 2.3. Ownership of the world fleet, as of 1 January 2015 (dwt)

Rank (dwt)	Country/territory of ownership	Number of vessels			Dead-weight tonnage				
		National flag	Foreign flag	Total	National flag	Foreign flag	Total	Foreign flag as a % of total	Total as a % of world
1	Greece	796	3 221	4 017	70 425 265	209 004 526	279 429 790	74.80%	16.11%
2	Japan	769	3 217	3 986	19 497 605	211 177 574	230 675 179	91.55%	13.30%
3	China	2 970	1 996	4 966	73 810 769	83 746 441	157 557 210	53.15%	9.08%
4	Germany	283	3 249	3 532	12 543 258	109 492 374	122 035 632	89.72%	7.04%
5	Singapore	1 336	1 020	2 356	48 983 688	35 038 564	84 022 252	41.70%	4.84%
6	Republic of Korea	775	843	1 618	16 032 807	64 148 678	80 181 485	80.00%	4.62%
7	Hong Kong, China	727	531	1 258	56 122 972	19 198 299	75 321 271	25.49%	4.34%
8	United States	789	1 183	1 972	8 731 781	51 531 743	60 263 524	85.51%	3.47%
9	United Kingdom	477	750	1 227	12 477 513	35 904 386	48 381 899	74.21%	2.79%
10	Norway	848	1 009	1 857	17 066 669	29 303 873	46 370 542	63.20%	2.67%
11	Taiwan Province of China	117	752	869	4 681 240	40 833 077	45 514 317	89.71%	2.62%
12	Bermuda	5	317	322	289 818	41 932 611	42 222 429	99.31%	2.43%
13	Denmark	392	538	930	15 286 153	20 893 511	36 179 664	57.75%	2.09%
14	Turkey	576	954	1 530	8 321 506	19 366 264	27 687 770	69.95%	1.60%
15	Monaco		260	260		23 929 323	23 929 323	100.00%	1.38%
16	Italy	596	207	803	15 961 983	6 040 199	22 002 182	27.45%	1.27%
17	India	697	147	844	14 546 706	7 268 449	21 815 155	33.32%	1.26%
18	Brazil	228	163	391	3 150 493	17 308 798	20 459 291	84.60%	1.18%
19	Belgium	87	156	243	7 302 545	12 787 196	20 089 741	63.65%	1.16%
20	Russian Federation	1 291	448	1 739	5 920 435	12 403 644	18 324 079	67.69%	1.06%
21	Islamic Republic of Iran	157	70	227	3 986 804	14 093 340	18 080 144	77.95%	1.04%
22	Switzerland	47	291	338	1 403 668	16 492 768	17 896 436	92.16%	1.03%
23	Indonesia	1 504	153	1 657	12 908 577	4 120 935	17 029 512	24.20%	0.98%
24	Netherlands	775	445	1 220	6 589 901	10 415 708	17 005 609	61.25%	0.98%
25	Malaysia	466	142	608	8 430 359	7 707 526	16 137 885	47.76%	0.93%
26	United Arab Emirates	95	684	779	472 967	14 845 550	15 318 518	96.91%	0.88%
27	Saudi Arabia	86	155	241	2 004 631	11 358 349	13 362 980	85.00%	0.77%
28	France	180	277	457	3 517 344	7 636 312	11 153 656	68.46%	0.64%
29	Cyprus	141	179	320	3 811 947	6 858 661	10 670 608	64.28%	0.62%
30	Viet Nam	786	92	878	6 527 639	1 510 645	8 038 284	18.79%	0.46%
31	Kuwait	42	27	69	5 293 213	2 462 656	7 755 869	31.75%	0.45%
32	Canada	209	139	348	2 743 006	5 004 054	7 747 060	64.59%	0.45%
33	Oman	6	31	37	5 842	7 008 489	7 014 331	99.92%	0.40%
34	Sweden	101	234	335	1 248 460	5 194 955	6 443 415	80.62%	0.37%
35	Qatar	56	70	126	888 093	5 471 554	6 359 647	86.04%	0.37%
	Total top 35 ship-owning countries	18 410	23 950	42 360	470 985 656	1 171 491 033	1 642 476 689	71.32%	94.69%
	All others	2 962	2 486	5 448	35 004 138	51 845 622	86 849 760	59.70%	5.01%
	Unknown country of ownership			717			5 234 918		0.30%
	WORLD TOTAL			48 525			1 734 561 367		100.00%

Source: UNCTAD secretariat, based on data supplied by Clarksons Research. For a complete listing of nationally owned fleets, see http://stats.unctad.org/fleetownership.

Note: Propelled seagoing vessels of 100 GT and above.

fleet ownership is Angola, followed by Nigeria and Egypt (see also the extended data available online for all ship-owning countries under UNCTADstat fleet ownership database – http://stats.unctad.org/fleetownership.

China, Indonesia and the Russian Federation have a large number of nationally flagged and owned ships, which are largely employed in coastal or inter-island shipping. These markets tend to be protected from foreign competition and do not necessarily fall under global IMO regulations. Ships deployed on these services tend to be smaller and older than the fleet deployed on international routes.

2. Container ship operators

Together, the three largest liner shipping companies, that is, those companies that operate the container ships deployed on regular services, have a share of almost 35 per cent of the world total container-carrying capacity. The top three companies are headquartered in Europe (Denmark, Switzerland and France), while most other carriers among the top 20 are based in Asia and one company in South America (Compañía Sudamericana de Vapores (CSAV)), headquartered in Santiago; the company has recently merged with Hapag Lloyd (headquartered in Germany) (table 2.4). Note that about half of the ships operated by the liner companies are not owned by them, but are chartered from the shipowner, who is likely to be from a third country, for example Germany or Greece.

Concentration in the sector continues to increase and the recent mergers of CSAV and Hapag Lloyd, and Compañía Chilena de Navegación Interoceánica and Hamburg Süd, have contributed further to this development. In the beginning of 2015, the top ten companies operated over 61 per cent of the global container fleet and the top 20 controlled 83 per cent of all capacity. All companies with vessels on order are investing in larger vessels, as the average vessel size of the order book is in all cases larger than the current average container-carrying capacity.

Table 2.4. The 50 leading liner companies, 1 May 2015 (Number of ships and total shipboard capacity deployed, ranked by TEU)

Rank	Operator	Market share % (TEU)	TEU	Vessels	Average vessel size	Orderbook TEU	Orderbook vessels	Average vessel size orderbook
1	Maersk Line A/S	13.45	2 526 490	478	5 286	91 080	9	10 120
2	Mediterranean Shipping Company (MSC) SA	13.22	2 483 979	451	5 508	498 680	36	13 852
3	CMA CGM S.A.	8.00	1 502 417	375	4 006	182 500	16	11 406
4	Evergreen Marine Corporation (Taiwan) Limited (Evergreen Line)	5.08	954 280	204	4 678	354 000	23	15 391
5	COSCO Container Lines Limited (COSCON)	4.55	854 171	158	5 406	119 500	10	11 950
6	China Shipping Container Lines Company Limited	4.00	751 507	136	5 526	19 100	1	19 100
7	Hapag-Lloyd Aktiengesellschaft	3.90	732 656	145	5 053	0	-	
8	Hanjin Shipping Company Limited	3.41	640 490	104	6 159	0	-	
9	Mitsui O.S.K. Lines Limited (MOL)	3.19	599 772	111	5 403	122 300	6	20 383
10	APL Limited	2.91	545 850	96	5 686	0	-	
11	Orient Overseas Container Line Limited (OOCL)	2.77	520 328	103	5 052	143 656	8	17 957
12	Hamburg Sudamerikanische Dampfschifffahrts-Gesellschaft KG	2.66	498 902	104	4 797	0	-	

Table 2.4. The 50 leading liner companies, 1 May 2015 (Number of ships and total shipboard capacity deployed, ranked by TEU) *(continued)*

Rank	Operator	Market share % (TEU)	TEU	Vessels	Average vessel size	Orderbook TEU	Orderbook vessels	Average vessel size orderbook
13	Nippon Yusen Kabushiki Kaisha (NYK)	2.63	494 953	104	4 759	112 000	8	14 000
14	Yang Ming Marine Transport Corporation	2.60	487 771	103	4 736	182 000	13	14 000
15	Hyundai Merchant Marine Company Limited (HMM)	2.13	399 791	65	6 151	60 000	6	10 000
16	Kawasaki Kisen Kaisha Limited ('K' Line)	2.12	397 623	77	5 164	110 960	8	13 870
17	Pacific International Lines (Private) Limited (PIL)	1.99	374 849	139	2 697	22 905	6	3 818
18	United Arab Shipping Company (S.A.G.) (UASC)	1.98	372 841	53	7 035	214 300	13	16 485
19	Zim Integrated Shipping Services Limited	1.58	296 554	66	4 493	0	-	
20	Compania Sud Americana de Vapores S.A. (CSAV)	1.26	237 567	40	5 939	18 000	2	9 000
21	Wan Hai Lines Limited	1.07	200 970	88	2 284	0	-	
22	X-Press Feeders	0.67	126 009	87	1 448	0	-	
23	MCC Transport (Singapore) Private Limited	0.58	109 662	62	1 769	0	-	
24	Delmas	0.53	99 078	47	2 108	0	-	
25	SITC Container Lines Company Limited	0.41	76 765	63	1 218	14 400	8	1 800
26	Korea Marine Transport Company Limited (KMTC Line)	0.40	75 779	46	1 647	5 400	1	5 400
27	Nile Dutch Africa Line BV	0.40	75 678	29	2 610	0	-	
28	United States Military Sealift Command	0.36	68 334	58	1 178	0	-	
29	Compania Chilena de Navegacion Interoceanica S.A. (CCNI)	0.32	59 906	14	4 279	18 030	2	9 015
30	CNC Line Limited	0.32	59 787	26	2 300	0	-	
31	BBC Chartering & Logistic GmbH & Company KG	0.31	57 570	93	619	0	-	
32	TS Lines Company Limited	0.31	57 477	36	1 597	0	-	
33	Safmarine Container Lines N.V.	0.28	52 638	23	2 289	0	-	
34	Arkas Konteyner ve Tasimacilik A.S.	0.28	52 180	36	1 449	5 000	2	2 500
35	Seago Line	0.27	50 688	22	2 304	0	-	
36	Simatech Shipping & Forwarding L.L.C.	0.24	45 947	19	2 418	8 700	2	4 350
37	Sinotrans Container Lines Company Limited (Sinolines)	0.23	43 447	36	1 207	16 000	4	4 000

Table 2.4. The 50 leading liner companies, 1 May 2015 (Number of ships and total shipboard capacity deployed, ranked by TEU) *(continued)*

Rank	Operator	Market share % (TEU)	TEU	Vessels	Average vessel size	Orderbook TEU	Orderbook vessels	Average vessel size orderbook
38	Regional Container Lines Public Company Limited	0.23	43 371	29	1 496	0	-	
39	ANL Singapore Private Limited	0.22	41 660	12	3 472	0	-	
40	Gold Star Line Limited	0.22	41 474	17	2 440	0	-	
41	Hafiz Darya Shipping Company (HDS Line)	0.22	41 337	9	4 593	0	-	
42	Grimaldi Group S.p.A.	0.21	40 262	41	982	0	-	
43	Unifeeder A/S	0.20	36 711	37	992	0	-	
44	Westfal-Larsen Shipping AS	0.19	35 151	17	2 068	0	-	
45	Swire Shipping Limited	0.18	34 333	24	1 431	0	-	
46	Seaboard Marine Limited	0.17	32 358	26	1 245	0	-	
47	Sinokor Merchant Marine Company Limited	0.17	31 969	32	999	0	-	
48	Spliethoff's Bevrachtingskantoor B.V.	0.17	31 454	36	874	0	-	
49	Heung-A Shipping Company Limited	0.17	31 332	31	1 011	5 400	3	1 800
50	Samudera Shipping Line Limited	0.16	30 995	26	1 192	3 600	2	1 800

Source: UNCTAD secretariat, based on data provided by Lloyd's List Intelligence.
Note: Includes all container-carrying ships known to be operated by liner shipping companies.

It is important to note that the attempt to realize economies of scale leads to new vessel orders that at the same time increase the risk of oversupply. The average vessel size for all new vessels on order by the top 15 companies is above 10,000 TEUs, which is double the current average size of vessels in the existing fleet of each company. Only very few companies outside the top 20 carriers have placed any new orders, and if at all, these orders are for far smaller vessel sizes.

The need to confront the oversupply has resulted in more frequent and wider cooperation of shipping lines on all routes, thus providing more and more homogenous services. A resulting challenge in the industry is the difficulty of service differentiation as container transport is a highly standardized transport service and shipping lines are rarely in a position to establish differentiation of services in terms of quality.

The trend towards larger ships, mergers and more collaboration is also reflected on individual routes and markets. The next section, on container ship fleet deployment, provides a more detailed analysis.

C. CONTAINER SHIP DEPLOYMENT AND LINER SHIPPING CONNECTIVITY

Since 2004, the UNCTAD LSCI has provided an indicator of each coastal country's access to the global liner shipping network, that is the network of regular maritime transport services for containerized cargo. The complete time series is published in electronic format on UNCTADstat (http://stats.unctad.org/lsci). The LSCI is generated from five components that capture the deployment of container ships by liner shipping companies to a country's ports of call: (a) the number of ships; (b) their total container-carrying capacity; (c) the number of companies providing services with their own operated ships; (d) the number of services provided; and (e) the size (in TEUs) of the largest ship deployed.

The country with the highest LSCI is China, followed by Singapore, Hong Kong (China), the Republic of Korea, Malaysia, and Germany. The best connected countries in Africa are Morocco, Egypt and South

Africa, reflecting their geographical position at the corners of the continent. In Latin America, Panama has the highest LSCI, benefiting from its canal and the location at the crossroads of main East–West and North–South routes, followed by Mexico, Colombia and Brazil. The 10 economies with the lowest LSCI are all island States, reflecting their low trade volumes and remoteness.

The LSCI of a country is not only determined by its trade volume, but increasingly by its position within the global liner shipping network. The relevance of hubs becomes evident in a high level of connectivity despite a relative low level of trade; examples are Jamaica, Morocco, Panama and Sri Lanka. The centrality of these countries in the global network is of high relevance for the regions in which they are located, as these points offer a high level of connectivity beyond the traditional direct connectivity.

Only 17–18 per cent of pairs of countries are connected with each other through a direct service; all other country pairs need to make use of at least one trans-shipment for bilateral containerized trade (Fugazza et al., 2013; Fugazza, 2015). Trans-shipment in many trade relations is growing and widely practiced in the industry to reach economies of scale and density in operations, and thus it is also widely accepted by customers as trans-shipment operations have become very efficient and the switch between services is often made in few hours.

Building on UNCTAD's newly developed LSBCI (http://stats.unctad.org/lsbci (accessed 15 July 2015), UNCTAD research suggests that lacking a direct maritime connection with a trade partner is associated with lower values of exports. Estimates point to a range varying from minus 42 per cent to minus 55 per cent. When assessing the effect of the number trans-shipments necessary to connect country pairs, any additional trans-shipment is associated with a 20–25 per cent lower value of exports. Results further suggest that in the absence of a bilateral connectivity indicator the impact of bilateral distance on bilateral exports is likely to be overestimated in statistical estimations (Fugazza, 2015).

A view of the level of connectivity from a bilateral perspective shows that intraregional routes are those with the greatest services capacity. The bilateral perspective further opens up the possibility of taking a closer look at the level of competition. It shows that on only 32 per cent of all 11,650 bilateral connections, including options with trans-shipments, are there five or more providers. Competition is limited on the remaining 68 per cent as the number of companies offering services is smaller or equal to four. This situation particularly affects small economies and island States. The most competitive routes for direct container shipping services are intraregional in Asia and Europe. There are 51 liner companies that have vessels deployed on routes that directly connect Singapore with ports in Malaysia, 46 companies provide direct services between China and the Republic of Korea, and 44 carriers offer a direct connection between the Netherlands and the United Kingdom (see table 2.5).

Table 2.5. Container ship deployment on selected routes, 1 May 2015

Direct services	Number of companies (vessel operators)	Largest vessel (TEU)
Malaysia – Singapore	51	15 908
China – Republic of Korea	46	19 224
Netherlands – United Kingdom	44	19 224
China – Japan	39	13 092
Germany – Netherlands	36	19 224
China – Singapore	35	15 908
Japan – Republic of Korea	35	10 000
Argentina – Brazil	23	9 700
China – United States	23	13 360
Panama – United States	21	5 116
China – Germany	19	19 224
Côte d'Ivoire – Nigeria	19	8 540
Chile – Peru	18	10 000
China – South Africa	16	10 000
United Republic of Tanzania – Mozambique	6	3 091
Kenya – Malaysia	5	3 108
Comoros – United Arab Emirates	3	2 226
Fiji – Australia	3	2 742
Dominica – United States	1	600
Japan – Marshall Islands	1	970

Source: UNCTAD secretariat, based on data supplied by Lloyd's List Intelligence.

Figure 2.6 further illustrates the process of concentration in liner shipping. While the container-carrying capacity per provider per country tripled between 2004 and 2015, the average number of companies that provide services to each country's ports decreased by 29 per cent. Both trends illustrate two sides of the same coin. As ships get bigger and companies aim at achieving economies of scale, there remain fewer companies in individual markets. It is a challenge for policymakers to support technological advances and cost savings, for example through economies of scale, yet at the same time ensure a sufficiently competitive environment so that cost savings are effectively passed on to the clients, that is, importers and exporters.

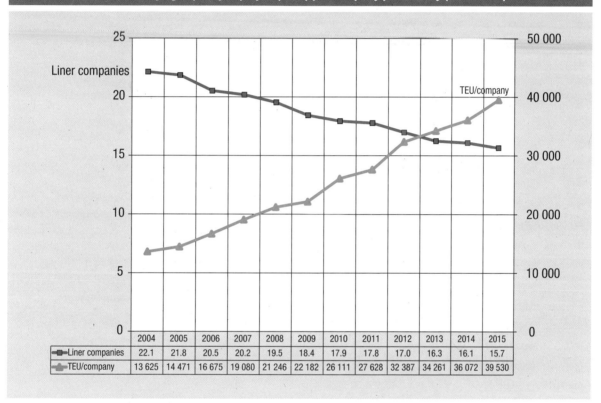

Figure 2.6. Presence of liner shipping companies: Average number of companies per country and average container-carrying capacity deployed (TEUs) per company per country (2004–2015)

	2004	2005	2006	2007	2008	2009	2010	2011	2012	2013	2014	2015
Liner companies	22.1	21.8	20.5	20.2	19.5	18.4	17.9	17.8	17.0	16.3	16.1	15.7
TEU/company	13 625	14 471	16 675	19 080	21 246	22 182	26 111	27 628	32 387	34 261	36 072	39 530

Source: UNCTAD secretariat, based on data supplied by Lloyd's List Intelligence.

D. REGISTRATION OF SHIPS

As of 1 January 2015, Panama, Liberia and the Marshall Islands are the largest vessel registries. Together, they account for a 41.8 per cent share of the world tonnage, with the Marshall Islands having recorded an impressive growth of over 13 per cent over 2014 (table 2.6). More than three quarters of the world fleet are registered in developing countries (table 2.7), including in many open registries, that is, registries where the owner does not need to be of the same nationality as the country where the ship is registered. The tonnage registered under a foreign flag (where the nationality of the owner is different from the flag flown by the vessel) is 71 per cent of the world total (see also table 2.3 above).

Care has to be taken when interpreting the data, as several registries have outsourced important parts of their operations and thus not all revenues remain in the flag State. Nevertheless, for some developing countries the provision of flag State services has become an important source of income.

Historically, when the first shipowners started to "flag out" by registering their ships in a foreign open registry in the 1970s or even earlier, one of the motivations may have been less stringent safety and environmental regulations. Today, there is no generalized difference between open and national registries as far as the ratification and implementation of relevant international conventions is concerned. A comparative table provided by the International Chamber of Shipping

Table 2.6. The 35 flags of registration with the largest registered fleets, as of 1 January 2015 (dwt)

Flag of registration	Number of vessels	Share of world total, vessels	Deadweight tonnage (1,000 dwt)	Share of world total (dwt)	Cumulated share (dwt)	Average vessel size (dwt)	Dwt growth 2015/2014 as %
Panama	8 351	9.33	352 192	20.13	20.13	44 052	0.91
Liberia	3 143	3.51	203 832	11.65	31.79	65 018	0.31
Marshall Islands	2 580	2.88	175 345	10.02	41.81	67 990	13.32
Hong Kong (China)	2 425	2.71	150 801	8.62	50.43	63 575	6.47
Singapore	3 689	4.12	115 022	6.58	57.01	33 830	8.52
Malta	1 895	2.12	82 002	4.69	61.70	43 898	8.69
Greece	1 484	1.66	78 728	4.50	66.20	63 286	4.45
Bahamas	1 421	1.59	75 779	4.33	70.53	54 322	2.54
China	3 941	4.41	75 676	4.33	74.85	20 756	-1.28
Cyprus	1 629	1.82	33 664	1.92	76.78	32 000	3.96
Isle of Man	1 079	1.21	23 008	1.32	78.09	55 441	-2.28
Japan	5 224	5.84	22 419	1.28	79.38	5 558	7.47
Norway	1 558	1.74	20 738	1.19	80.56	15 339	-1.20
Italy	1 418	1.58	17 555	1.00	81.57	14 556	-11.22
United Kingdom	1 865	2.08	17 103	0.98	82.54	16 059	-0.35
Republic of Korea	673	0.75	16 825	0.96	83.51	10 099	-3.13
Denmark	7 373	8.24	16 656	0.95	84.46	26 606	13.94
Indonesia	1 604	1.79	15 741	0.90	85.36	3 681	2.29
India	1 174	1.31	15 551	0.89	86.25	10 157	-1.39
Antigua and Barbuda	650	0.73	12 753	0.73	86.98	10 909	-3.45
Germany	3 561	3.98	12 693	0.73	87.70	22 230	-11.69
United States	1 613	1.80	12 683	0.73	88.43	6 089	2.59
United Republic of Tanzania	1 313	1.47	11 703	0.67	89.10	46 256	-1.54
Bermuda	1 245	1.39	11 511	0.66	89.75	71 946	2.69
Malaysia	1 777	1.99	9 232	0.53	90.28	6 793	-0.95
Turkey	2 471	2.76	8 820	0.50	90.79	8 181	-2.64
Netherlands	1 412	1.58	8 651	0.49	91.28	7 536	0.34
Belgium	756	0.85	8 609	0.49	91.77	45 548	21.96
Viet Nam	674	0.75	7 351	0.42	92.19	4 499	0.81
Russian Federation	963	1.08	7 221	0.41	92.60	2 974	2.45
France	670	0.75	6 882	0.39	93.00	16 042	-8.85
Philippines	646	0.72	6 850	0.39	93.39	6 149	6.19
Kuwait	765	0.86	5 440	0.31	93.70	40 002	37.91
Thailand	749	0.84	5 070	0.29	93.99	7 636	0.86
Taiwan Province of China	586	0.66	4 829	0.28	94.27	18 431	8.05
Top 35 total	72 377	80.90	1 648 937	94.27	94.27	27 697	3.53
World total	89 464	100.00	1 749 222	100.00	100.00	22 757	3.54

Source: UNCTAD secretariat, based on data supplied by Clarksons Research.
Note: Propelled seagoing merchant vessels of 100 GT and above, ranked by dead-weight tonnage. For a complete list of all countries see http://stats.unctad.org/fleet (accessed 19 September 2015).

CHAPTER 2: STRUCTURE, OWNERSHIP AND REGISTRATION OF THE WORLD FLEET

Table 2.7. Distribution of dwt capacity of vessel types, by country group of registration, January 2015 (beginning-of-year figures, per cent of dwt, annual growth in percentage points in italics)

	Total fleet	Oil tankers	Bulk carriers	General cargo	Container ships	Others
World total	100.00	100.00	100.00	100.00	100.00	100.00
Developed countries	22.70	26.26	17.82	28.38	26.81	25.75
	-0.02	*-0.09*	*-0.05*	*-0.02*	*0.54*	*-0.08*
Countries with economies	0.71	0.78	0.26	5.35	0.03	1.22
in transition	*0.00*	*0.01*	*0.01*	*-0.03*	*0.00*	*0.01*
Developing countries	76.36	72.91	81.90	65.41	73.14	71.45
	0.03	*0.08*	*0.06*	*-0.05*	*-0.55*	*0.05*
Of which:						
Africa	13.14	17.18	9.98	5.96	20.19	9.93
	-0.46	*-0.25*	*-0.44*	*0.06*	*-1.11*	*-0.51*
America	26.74	20.68	31.93	22.57	19.75	31.53
	-0.68	*-0.24*	*-0.63*	*-0.76*	*-2.24*	*-0.66*
Asia	26.05	21.46	29.46	33.92	28.00	18.92
	0.27	*-0.07*	*-0.10*	*0.67*	*2.27*	*-0.08*
Oceania	10.42	13.60	10.53	2.95	5.20	11.07
	0.85	*1.10*	*0.87*	*0.03*	*0.74*	*0.76*
Unknown and other	0.24	0.05	0.01	0.86	0.02	1.57
	0.00	*0.00*	*-0.03*	*0.10*	*0.01*	*0.02*

Source: UNCTAD secretariat, based on data supplied by Clarksons Research.
Note: Propelled seagoing merchant vessels of 100 GT and above.

shows that both national and open registries can be found among the best and among the worst service providers (International Chamber of Shipping, 2014). The registries with youngest fleets among the top 35 flags were Hong Kong (China), the Marshall Islands and Singapore.

Registries with a good track record usually host far younger fleets and keep a close eye on the compliance of shipowners with international regulations. It is in their interest that their flag is not targeted by port State control authorities, as this would make the flag less attractive to shipowners. It is, in fact, in the interest of these "good" registries that environmental and safety regulations are ambitious and strictly enforced, as this will be more of a challenge to owners and registries with older and less well-maintained ships.

Interestingly, several of the major open registries are located in SIDS. These registries have a double interest in promoting ambitious regulations, for example within IMO. If, for example, lower global limits on CO_2 emissions are imposed, this could further enhance the competitive advantage of those registries that already host more modern and younger fleets. It would also constitute a contribution to climate change mitigation, which is of paramount concern for many island economies.

E. SHIPBUILDING, DEMOLITION AND NEW ORDERS

1. Deliveries of newbuildings

In total, the world fleet grew by 42 million GT in 2014, resulting from newbuildings of almost 64 million GT minus reported demolitions of about 22 million GT.

More than 91 per cent of GT delivered in 2014 was built in just three countries: China (35.9 per cent); the Republic of Korea (34.4 per cent); and Japan (21.0 per cent), with China mostly building dry bulk carriers, followed by container ships and tankers; the Republic of Korea building mostly container ships and oil tankers; and Japan specializing fundamentally in bulk carriers.

To respond to demands for a more environmentally sustainable shipping fleet, shipbuilders, owners and non-governmental technical bodies such as classification societies increasingly collaborate on the development of new technologies and eco-ships. Notably, classification societies have in recent years led research into the use of alternative energies on ships, including wind and solar power.

Table 2.8.	Deliveries of newbuildings, major vessel types and countries where built (2014, thousands of GT)					
	China	Republic of Korea	Japan	Philippines	Rest of world	World total
Oil tankers	2 896	4 781	891		466	9 034
Bulk carriers	13 304	1 588	10 791	869	167	26 719
General cargo	585	329	199		372	1 485
Container ships	4 986	9 135	188	995	735	16 039
Gas carriers	119	3 528	666		14	4 328
Chemical tankers	113	185	188		57	543
Offshore	714	1 485	51		956	3 206
Ferries and passenger ships	92	5	27		767	892
Other	42	835	391		147	1 415
Total	22 851	21 872	13 392	1 865	3 682	63 662

Source: UNCTAD secretariat, based on data provided by Clarksons Research.
Note: Propelled seagoing merchant vessels of 100 GT and above. More detailed data on other countries where vessels were built is available under http://stats.unctad.org/shipbuilding.

2. Demolition of ships

The scrapping of ships helps reduce the oversupply of tonnage, and it encourages the modernization of the fleet, including from an ecological perspective, as the vessels demolished tend to be less fuel efficient and more detrimental to the environment as far as emissions are concerned. The Government of China has extended a subsidy programme that encourages shipping companies to scrap old vessels. The scheme, which began in 2013, provides financial incentives to shipowners to replace old vessels with newer, more environmentally friendly models (Reuters, 2015).

Table 2.9.	Tonnage reported sold for demolition, major vessel types and countries where demolished (2014, thousands of GT)							
	India	China	Bangladesh	Pakistan	Turkey	Unknown Indian subcontinent	Others/ unknown	World total
Oil tankers	393	827	368	2 227	86	160	420	4 482
Bulk carriers	1 576	2 771	2 888	1 458	151	111	143	9 098
General cargo	719	301	313	65	349		259	2 008
Container ships	3 455	777	303	32	63		139	4 769
Gas carriers	215	8	62		28		29	342
Chemical tankers	136	3	10	13	34		1	196
Offshore	127	6	199	331	9		26	697
Ferries and passenger ships	74	13	19		67		22	194
Other	270	168	106		53		12	609
Total	6 965	4 873	4 269	4 127	839	271	1 051	22 394

Source: UNCTAD secretariat, based on data from Clarksons Research.
Note: Propelled seagoing merchant vessels of 100 GT and above. More detailed data on other countries where vessels were demolished is available under http://stats.unctad.org/shipscrapping.

South Asia (Bangladesh, India and Pakistan) and China together account for more than 90 per cent of global ship breaking. Within ship demolition, furthermore, a certain specialization exists, as most container ships are demolished in India, while Bangladesh and China purchased more dry bulk carriers, and Pakistan mostly oil tankers.

Ship breaking itself is also under scrutiny for its environmental impact, particularly the method of "beaching" applied in South Asia, which tends to be harmful to the local environment and often lacks health and safety measures. Ongoing projects aim at the development of safe and environmentally sound ship recycling, with the goal of improving the standards and, therefore, the sustainability of the industry (IMO, 2015).

3. Tonnage on order

The world order book in early 2015 is far below its peak of 2008–2009. Between 2014 and 2015, the order book declined for most vessel types except for oil tankers. Those who did place new orders did so for two main reasons: first, they expect future demand to grow sufficiently to cater for the new deliveries; second, new ships are more fuel efficient and less polluting. To comply with new regulations having as objective the long-term environmental sustainability of international shipping, shipowners find additional motivations to replace old tonnage with newbuildings. In April 2015, the container ship order book stood at 18 per cent of existing capacity, its lowest level for over a decade (Clarksons Research, 2015b).

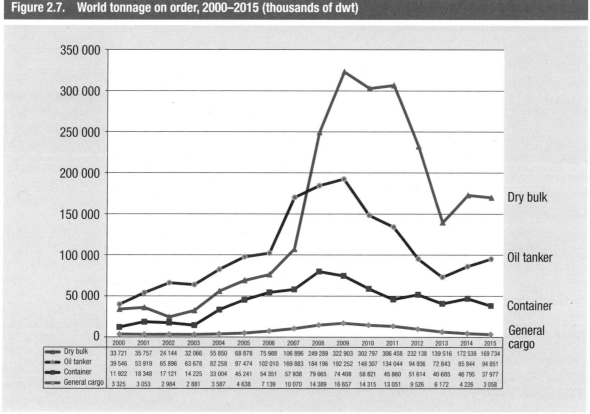

Figure 2.7. World tonnage on order, 2000–2015 (thousands of dwt)

Source: UNCTAD secretariat, based on data supplied by Clarksons Research.
Note: Propelled seagoing merchant vessels of 100 GT and above, beginning of year figures.

4. Outlook

Economic and regulatory incentives will continue to encourage individual owners to invest in modernizing their fleets. Unless older tonnage is demolished, this would lead to further global overcapacity, continuing the downward pressure on freight and charter rates (see also chapter 3). The interplay between more stringent environmental regulations and low freight and charter rates should encourage the further demolition of older vessels; this would not only help reduce the oversupply in the market, but also contribute to lowering the global environmental impact of shipping.

REFERENCES

Clarksons Research (2014a). *World Fleet Monitor*. January.

Clarksons Research (2014b). *World Fleet Monitor*. November.

Clarksons Research (2015a). *World Fleet Monitor*. March.

Clarksons Research (2015b). *Container Intelligence Monthly*. April.

Fugazza M (2015). Maritime connectivity and trade. UNCTAD Policy Issues in International Trade and Commodities No. 70. Geneva.

Fugazza M, Hoffmann J and Razafinombana R (2013). Building a data set for bilateral maritime connectivity. UNCTAD Policy Issues in International Trade and Commodities No. 61. Geneva.

IMO (2015): Recycling of ships. Available at http://www.imo.org/en/OurWork/Environment/ShipRecycling/Pages/Default.aspx (accessed 13 July 2015).

International Chamber of Shipping (2014). Shipping industry flag State performance table 2014–2015. Available at http://www.ics-shipping.org/docs/default-source/resources/policy-tools/ics-shipping-industry-flag-state-performance-table-2014-15.pdf?sfvrsn=10 (accessed 13 July 2015).

Lloyd's List Intelligence – Containers (2015). Available at http://www.lloydslistintelligence.com/llint/containers/index.htm (accessed 1 May 2015).

Morgan Stanley (2013). Maritime industries. Eco ships: Fuel savings alone do not justify investing in newbuilds. Morgan Stanley Research. 26 August.

Reuters (2015). UPDATE 1 – China extends ship scrapping subsidy programme to end-2017. Industries. 23 June. Available at http://www.reuters.com/article/2015/06/23/china-shipping-idUSL3N0Z91FB20150623 (accessed 28 June 2015).UNCTADstat – Fleet ownership. See http://stats.unctad.org/fleetownership (accessed 1 July 2015).

UNCTADstat – Fleet ownership. See http://stats.unctad.org/fleetownership (accessed 1 July 2015).

ENDNOTES

[1] The underlying data on the world fleet for chapter 2 has been provided by Clarksons Research, London. The vessels covered in UNCTAD's analysis include all propelled seagoing merchant vessels of 100 GT and above, including offshore drillships, FPSOs and the Great Lakes fleets of Canada and the United States, which for historical reasons had been excluded in earlier issues of the *Review of Maritime Transport*. Military vessels, yachts, waterway vessels, fishing vessels and offshore fixed and mobile platforms and barges are excluded. As regards the main vessel types (oil tankers, dry bulk, container and general cargo carriers), there is no change compared to previous issues of the *Review of Maritime Transport*. As regards "other" vessels, the new data include a smaller number of ships (previously, fishing vessels with little cargo-carrying capacity had been included) and a slightly higher tonnage due to the inclusion of ships used in offshore transport and storage. To ensure full comparability of 2015 data with the four previous years, UNCTAD has updated the fleet data available online for the years 2011 to 2015, applying the same criteria (see http://stats.unctad.org/fleet). As in previous years, the data on fleet ownership covers only ships of 1,000 GT and above, as information on the true ownership is often not available for smaller ships. For more detailed data on fleet ownership see http://stats.unctad.org/fleetownership.

3

FREIGHT RATES AND MARITIME TRANSPORT COSTS

Developing countries, especially in Africa and Oceania, pay 40 to 70 per cent more on average for the international transport of their imports than developed countries. The main reasons for this situation are to be found in these regions' trade imbalances, pending port and trade facilitation reforms, as well as lower trade volumes and shipping connectivity. There is potential for policymakers to partly remedy the situation through investments and reforms, especially in the regions' seaports, transit systems and customs administrations.

Container freight rates remained volatile throughout 2014 although with different trends on individual trade lanes. Market fundamentals have not changed significantly despite the expansion in global demand for container shipping. This was mainly due to pressure from the constant supply of vessels that the market rates continued to face, with the introduction of very large units on mainlane trades and the cascading effect on non-mainlanes trades. The tanker market, which encompasses the transportation of crude oil, refined petroleum products and chemicals, witnessed an equally volatile freight rate environment in 2014 and early 2015. The dry bulk market freight rates faced another challenging year influenced by the surplus capacity that still exists and the uncertainties in demand projections. Bulk carrier earnings fell 5 per cent from 2013 to reach an average of $9,881 per day in 2014. The low level of earnings exerted financial pressure on owners and led to several companies filing for bankruptcy.

A. DETERMINANTS OF MARITIME TRANSPORT COSTS

Policymakers and shippers have an interest in understanding the determinants of international maritime transport costs. Maritime transport handles over 80 per cent of the volume of global trade (and about 90 per cent of developing countries' volume of international trade is seaborne) and knowing the reasons for differences in what a trader pays for the international transport of merchandise goods can help identify possible areas for intervention by policymakers. Extensive recent research has helped identify the main determinants of freight costs (see Cullinane et al., 2012; ECLAC, 2002; Sourdin and Pomfret, 2012; and Wilmsmeier, 2014; and the literature reviewed therein).

Figure 3.1 summarizes seven groups of determinants. The remainder of this section will introduce each one of these groups and discuss the options for policymakers to help reduce international maritime transport costs.

In recent years, policymakers and industry players have increasingly mainstreamed environmental sustainability criteria into their planning processes, policies and structures, not only to respond to global challenges for reducing emissions and improving the environmental footprint but also as a means to improve energy savings and to achieve a more efficient allocation of available resources. Specific actions may involve developing fuel-efficient vessels, improving energy efficiency, reshaping transport architecture and networks, adapting and developing appropriate infrastructure, rethinking and optimizing operating procedures of freight logistics, harnessing new technologies, and supporting information and communications technology and intelligent transport systems.

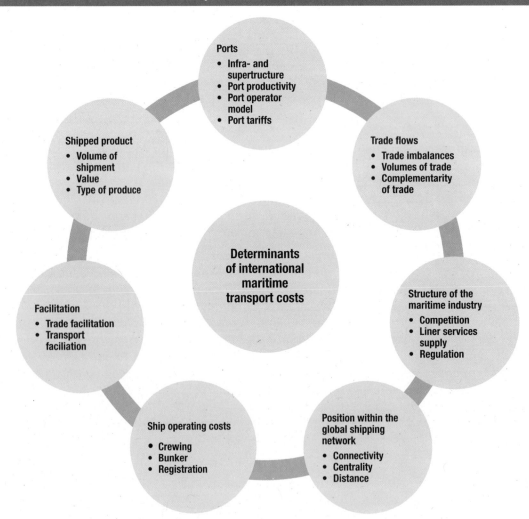

Figure 3.1. Determinants of maritime transport costs

Source: UNCTAD secretariat, based on Wilmsmeier, 2014.

1. Trade and transport facilitation

Reducing waiting times in seaports for ships and their cargo has a direct bearing on trade costs. First, from the shippers' perspective, it implies lower costs associated with the holding of inventory en route to the final destination. It has been estimated that each additional day cargo spends in transit is equivalent to an ad valorem tariff of 0.6 to 2.1 per cent (Hummels and Schaur, 2013). Second, waiting times also imply costs to the carrier, which will ultimately have to be passed on to the client through higher freight charges. Wilmsmeier et al. (2006) estimated that a 10 per cent reduction of the time it takes to clear customs implies a reduction of the maritime freight of about 0.5 per cent.

Different trade facilitation measures can be implemented to reduce waiting times and improve the logistics performance of countries in other ways. It has been suggested by UNCTAD (2015) that the transparent publication of trade-related information (such as measures included in article 1 of the WTO TFA) as well as the simplification and reduction of customs formalities (such as measures included in article 10 of the WTO TFA) have a particularly high statistical correlation with a country's ranking in international logistics benchmarks, such as the World Bank Logistics Performance Index (figure 3.2).

2. Ship operating costs

Technological advances have led to a continuous reduction in vessel operating costs over the decades. Improved fuel efficiency, economies of scale, and automation in port operations all help to reduce environmental and financial costs (see chapter 2).

However, the drive to invest in lower operating costs may have some negative repercussion on freight rates. For example, as carriers invest in larger and more energy-efficient vessels in the current market situation – to achieve economies of scale or to improve fuel efficiency – they inadvertently also contribute to a further oversupply of capacity. While the individual carrier may benefit from cost savings from deploying bigger vessels, all carriers bear the burden of the resulting oversupply and lower freight levels – to the benefit of importers and exporters.

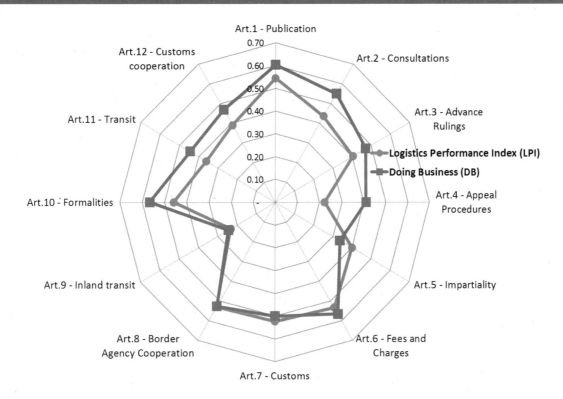

Figure 3.2. Statistical correlation between articles of the WTO TFA and indicators for trade efficiency

Source: UNCTAD secretariat, based on data from the World Bank (Logistics Performance Index and Doing Business Index) and WTO (number of category A notifications).

Note: The axis in the chart represents the partial correlation coefficient between the notification of trade facilitation measures under the 12 TFA articles and the value in the Logistics Performance and the Doing Business indices.

Oversupply of shipping capacity combined with a weak global economy has been a main factor affecting freight rates in recent years. In an effort to deal with low freight rate levels and to leverage some earnings, carriers have looked at measures to improve efficiency and optimize operations in order to reduce unit operating costs. Some of these measures involved operational consolidation, slow steaming, idling, and replacing smaller and older vessels with newer and more fuel efficient ones.

Although operating costs in shipping have been decreasing, the total costs of the transport system have declined less. First, total costs for the carrier have to take into account the costs of investing in new assets. Second, larger ships and the increasing use of hub ports also require ports and port cities to invest in additional capacities for storage, handling and intermodal connections. These additional costs – including external social and environmental costs – are not born by the carrier, but by the ports and local communities.

Lower operating costs as compared to higher fixed costs (that is, the capital costs associated with larger and more fuel-efficient ships) will likely also lead to more volatile freight rates. In the short term, the freight costs will have to cover at least the operating costs of the carrier; put differently, if the price of a transport service does not cover at least the fuel, communications and crewing costs, the carrier will anchor the ship and not offer the transport service. In the long term, however, the freight charges will have to cover the total average costs, including the fixed costs. As operating (variable) costs are lower today than in previous decades, this means that freight rates may also reach lower levels than in the past. Lower unit operating costs in bigger vessels, however, can only be reached if utilization rates are sufficient; if they are not, the carrier might be affected by diseconomies of scale. The risk of the latter also increases with ship size, particularly if demand and supply do not develop in line with each other. Effectively, freight rates appear to fluctuate more today than in earlier decades, and the changing structure of operating versus fixed costs is probably one of the reasons for this trend.

3. Distance and a country's position within shipping networks

Shipping goods over a longer distance requires more time (capital costs) and fuel (operating costs). Thus,

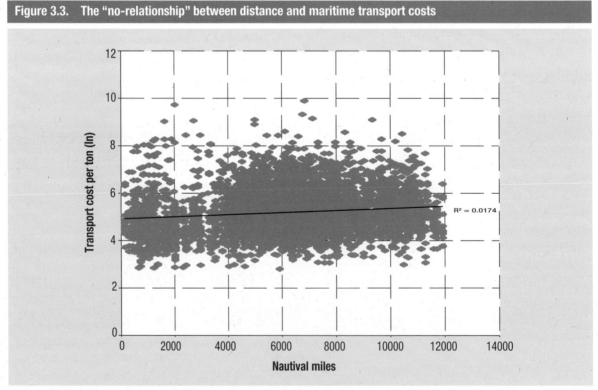

Figure 3.3. The "no-relationship" between distance and maritime transport costs

Source: ECLAC and UNCTAD secretariat, based on data from the International Transport Database – ECLAC, 2013.
Note: Based on 12,595 observations of maritime transport costs in international trade for the year 2013 at the Standard International Trade Classification two-digit level.

trading partners that are further away from main markets might expect to be also confronted with higher bilateral freight costs. As regards the impact of distance, the traditional gravity model would suggest that countries that are further away from each other will trade less (see, for example, Tinbergen, 1962; Pöyhönen, 1963; and Linnemann, 1966). However, traditional gravity models ignore effective distance and connectivity as potentially described by network structures (for example, the regular shipping liner services configuration). Limão and Venables (2001) show, using the example of shipping costs to Baltimore, that geographic distance alone cannot explain price differences in freight rates (figure 3.3).

Figure 3.3 illustrates that the geographical maritime distance only has a small statistical correlation with freight costs. More than the geographical distance, it may be rather the economical distance, as for example captured by shipping connectivity and a country's position within global shipping networks, that emerges as the relevant factor for international transport costs. Bilateral liner shipping connectivity, as captured by the UNCTAD LSBCI (see chapter 2) has a stronger bearing on freight costs than distance (figure 3.4).

Research on liner shipping connectivity frequently concludes that the position within a network has a more significant impact than the notion of geographical distance (Kumar and Hoffmann, 2002; Márquez-Ramos et al., 2005; Wilmsmeier et al., 2006; Wilmsmeier, 2014; Angeloudis et al., 2006; and McCalla et al., 2005). This important finding also needs to be seen in the context of the influencing variables of liner network connectivity such as ship size and frequency, which are determined by the overall level of trade, the geographic position and last but not least port infrastructure endowment and development options (see chapter 4).

The functioning of the network and its structure involve complex interaction between the maritime and port industry, and also the country and international organizations acting as governing and regulating bodies. Decisions made by these actors will subsequently also influence the cost of transport for a country or region in trade with its counterparts. Figure 3.4 (section C.3) exemplifies the reduction in freight rates with increasing connectivity, where connectivity is an expression of shipping possibilities, port infrastructure endowment and industry structure (for a detailed discussion, see Wilmsmeier and Hoffmann, 2008; and Wilmsmeier, 2014).

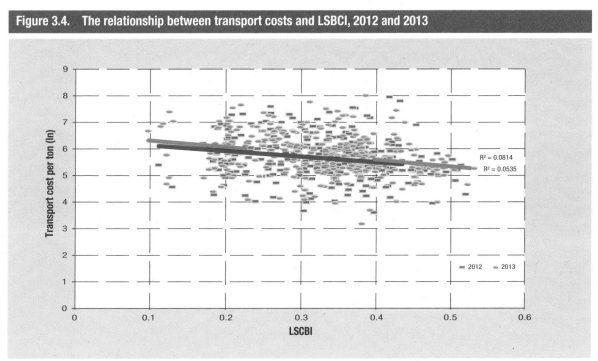

Figure 3.4. The relationship between transport costs and LSBCI, 2012 and 2013

Source: UNCTAD secretariat, based on data from the International Transport Database – ECLAC, 2012 and 2013.
Note: Based on 7,868 observations of maritime transport costs in international trade for the years 2012 and 2013 at the Standard International Trade Classification one-digit level.

4. Competition and market regulation

Price-setting in transport and logistics markets significantly depends on the level of effective competition. Competition in the transport markets depends on the size of the market and effective market regulation. Any impediment to free competition and the potential existence of collusive behaviour, atomization and monopolies will have impacts on price structures, and these factors are discussed in the following paragraphs.

Historically, shipping lines have tried to concentrate activities in accordance with other market players at certain points, as they are aware of the benefits of economies of agglomeration and scope. This has given room for the development of hub-and-spoke strategies and share capacity, in which the hubs are nodes for high-volume services to interchange cargoes and to transfer cargo to secondary routes.

The different strategies of shipping lines, the balance of power between shipping lines, shippers and ports, and constraints related to inland transportation can impact on the evolution and characteristics of and competition in maritime shipping networks. Moreover, strategic alliances between the port and the shipping industry, which have both been driven by strong concentration processes and vertical integration at global level, have a profound influence on maritime network structure and also on the degree of integration of a region in the global maritime transport network.

Policymakers need to carefully observe concentration processes in the maritime industry and be aware of possible negative effects on the trade and competitiveness of a country's exports, predominantly in network peripheral countries and regions. See figure 2.6 (chapter 2), which illustrates the decreasing number of shipping companies providing services in individual markets.

5. Value, volume and type of shipped product

The influence of the unitary value of the product on ocean freight rates has to be interpreted in the context of the history and structure of shipping markets. The value of the product also determines the elasticity of demand, that is, the willingness of the shipper to pay higher or even premium rates. Earlier works (Wilmsmeier, 2003; Wilmsmeier et

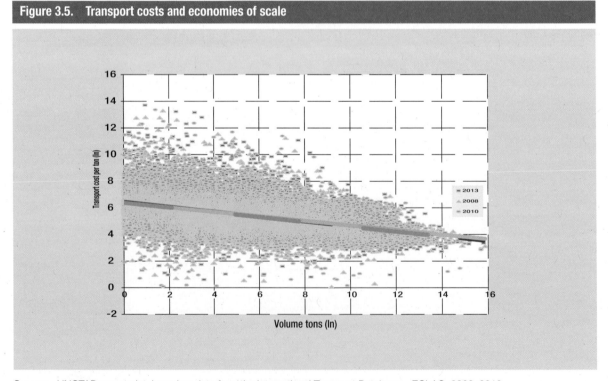

Figure 3.5. Transport costs and economies of scale

Source: UNCTAD secretariat, based on data from the International Transport Database – ECLAC, 2008–2013.
Note: All data are at the Standard International Trade Classification two-digit level, excluding products in Standard International Trade Classification commodity groups three and nine.

al., 2006; Martínez-Zarzoso and Suárez Burguet, 2005; and Wilmsmeier and Martínez-Zarzoso, 2010) all identify a relevance of the product unit value on transport costs. Palander (1935) had already proposed that transport costs were not regular but varied according to the weight, bulk, value and perishability of the product, and mode of transport and distance. Radelet and Sachs (1998) found that countries differed in their average "cost, insurance, freight"/"free on board" ratios not only due to differences in shipping costs but also due to differences in composition of commodity mix in external trade.

Despite the fact that there is no obvious reason for the connection between the freight rate and value of a product, a wide range of works describe the relationship between a product's unit value and the freight charged. The reason is that operators assume that unit value is inversely related to the elasticity of demand for transport. Besides insurance costs, feedering in hub-and-spoke networks, modal switching and the like, can also have an influence. Each product has a certain risk sensibility during transport. Risk in this context can refer to timely delivery, the probability of theft and/or high sensitivity to changes in the environment (temperature and the like).

Wilmsmeier and Sánchez (2009) analysed transport cost determinants for containerized food imports to South America and showed that a 10 per cent rise in the value of the commodity increased transport costs by around 7.6 percent. Special transport conditions and needs for certain types of cargo are also reflected in the structure of international maritime transport costs. Containerization has produced standard units in terms of size; nevertheless the requirements for transporting goods vary and thus different types of containers exist to satisfy these demands. The transport of refrigerated cargo has certain implications.

Economies of scale occur at two different levels. First, system internal economies of scale, which reflect the decrease in transport costs per ton, as the size of the individual shipment increases. Second, system external economies of scale, which reflect the decrease in transport costs as the volume of trade between two countries increases. The latter is also linked to other determinants of transport costs, such as levels of competition, vessel operation costs and port infrastructure.

Figure 3.5 illustrates the effect of economies of scale as volume per shipment. These are economies of scale realized outside a company as a result of its location and occur when trade between two countries has low tariffs and customs restrictions, or a region has an efficient and effective transport infrastructure.

6. Port characteristics and infrastructure

Port performance is essential for the efficiency and effectiveness of the maritime network. Port infrastructure endowment can be described by variables such as number of cranes, maximum draught and storage area at origin and destination ports. The interaction of these variables is decisive. Installing ship-to-shore gantries, for example, may well lead to higher port charges for the shipping line. The line may still achieve an overall saving, because its ships spend less time in the port, or because it can change from geared to gearless vessels. This, in turn, will also lead to lower freight rates.

However, development of port infrastructure is only worthwhile if the entire transport system benefits and not if bottlenecks are only shifted to another element within the system. Factors influencing productivity are physical, institutional and organizational. Physical limiting factors include the area, shape and layout of the terminal, the amount and type of equipment available, and the type and characteristics of the vessels using the terminal. Lack of cranes, insufficient land, oddly shaped container yards, inadequate berthage, inadequate gate facilities, and difficult road access are all physical limiting factors. Productivity must be considered in a system perspective for it to be of maximum value to industry. This is important from a policy perspective, thus emphasizing the need for co-modality and multimodal visions in policy recommendations and guidance. All players should have an awareness of the entire system and be wary of becoming its weak link.

Empirical results presented by Wilmsmeier et al. (2006) are quite clear and straightforward: increases in port efficiency, port infrastructure, private sector participation and inter-port connectivity all help to reduce the overall international maritime transport costs. If the two countries in their sample with the lowest port efficiency improved their efficiency to the level of the two countries with the highest indexes, the freight on the route between them would be expected to decrease by around 25 per cent. Improvements in port infrastructure and private sector participation, too, lead to reduced maritime transport costs. Unlike distance, port efficiency can be influenced by policymakers. Doubling port efficiency at both ends has the same effect on international maritime transport costs as would a "move" of the two ports 50 per cent closer to each other.

Hence, improving port efficiency and productivity and introducing technical advances as well as port design and planning measures to improve efficiencies and reduce time can reveal important insights to policymakers.

7. Trade flows and imbalances

The volume and type of cargo has a direct bearing on the carrier's costs. The volume of cargo is important as it allows for economies of scale, both on the sea leg as well as in port, although at times the economies of scale achieved on the shipping side may lead to congestion and diseconomies of scale in the port.

The extent to which the costs incurred by the carrier are passed on to the client depend on the market structure and also on the trade balance. On many shipping routes, especially for most bulk cargoes, ships sail full in one direction and return almost empty in the other. Having spare capacity, carriers are willing to transport cargo at a much lower freight rate than when the ships are already full. Freight rates are thus far higher from China to North America than for North American exports to China. By the same token, freight rates for containerized imports into Africa are higher than for exports. To some extent the differences in freight rates that depend on the direction of trade may be considered, in order that a market mechanism may help reduce imbalances. Those that have a trade deficit pay less for the transport of their exports.

In containerized trade, balance of trade flows is key in price-setting for shipping lines. Shipping lines calculate the costs to move a container on a return-trip basis, taking probability for empty positioning into account. When trade balance is negative, a country's imports exceed its exports and the greater the imbalance, the lower the freight rates will be for the country's exports; but if exports exceed imports, then the larger the imbalance, the higher the expected freight rates for exports will be. This divergence, associated with the sign of trade imbalance, occurs as a result of the freight rate price-fixing mechanisms applying in the liner market. Liner companies know that recurrently on one of the legs of the turnaround trip, the percentage of vessel capacity utilization will be lower, and therefore adapt the pricing scheme to the direction of the trip and to its corresponding expected cargo. Freight rates will be higher for the shipments transported on the leg of the trip with more traffic, as the total amount charged for this leg must compensate the relatively reduced income from the return trip, when part of the vessel's capacity will inevitably be taken up with repositioned empty containers. Excess capacity on the return trip will increase the competition between the various liner services, and as a result freight rates will tend to be lower. Organization of the transport service market can reduce empty movements through information and equipment sharing, freight-pooling, and transnational cooperation of transport service providers.

B. INTERNATIONAL TRANSPORT COSTS

International transport costs are a key component of trade costs and economic development. Recent research in Asia and the Pacific suggests that tariffs account for only 0–10 per cent of bilateral comprehensive trade costs, while other policy-related trade costs (that is, of a non-tariff nature) account for 60–90 per cent of bilateral trade costs. Put differently, issues such as transport costs, maritime connectivity, and procedures have a stronger bearing on trade costs than customs duties (Economic and Social Commission for Asia and the Pacific, 2015).

Based on data from merchandise imports, UNCTAD has estimated the expenditures on international transport (all modes of transport) for country groups (figure 3.6). For the average country, international transport costs amounted to approximately 9 per cent of the value of imports during the decade 2005–2014. Among the main regional groupings, African countries paid the most (average of 11.4 per cent) against an average of only 6.8 per cent for the developed countries.

Having considered the seven main determinants of maritime transport costs, it is now possible to discuss possible reasons for the overall freight costs estimated for different country groups, and in particular why Africa and Oceania pay more for the transport of their imports than other regions. These points are highlighted in the following paragraphs:

(a) *Trade and transport facilitation*: Many countries in Africa are landlocked, depending not only

CHAPTER 3: FREIGHT RATES AND MARITIME TRANSPORT COSTS

on the procedures of their own customs and other border agencies, but also on those of the neighbouring transit countries. This situation had been termed the "landlocked with bad neighbours trap" by Collier (2008). Many countries in Africa and Oceania also report low scores in indicators such as the Doing Business Index or the Logistics Performance Index (see chapter 5 on the linkages between trade facilitation and sustainable development).

(b) *Operating costs*: Operating costs (including costs of repairs and maintenance, hull and machinery and protection and indemnity insurance premiums, crewing, provisions, stores, water and lubricating oil) for vessel operators are overall the same, largely independent of routes or locations. However, these vary depending on ship type and age. Routes with bigger and/or newer vessels will have lower operating costs. Further operating costs may vary over time and depend on fuel prices, but they cannot systematically explain why freight rates would be higher on one route compared to another.

(c) *Position within shipping networks*: In particular, SIDS in Oceania are negatively affected by their geographical position, far from most major shipping routes. Promoting inter-island connectivity and developing regional/subregional hub ports, as well as upgrading or redeveloping port infrastructure and improving cargo handling with a potential to reduce freight costs, could be considered. In Africa, some countries have been able to benefit from their geographical position and offer trans-shipment services. Egypt, for example, benefits from the traffic passing through the Suez Canal, and Mauritius and Morocco both have established important hub ports. Most other African countries, however, are relatively far from the major East–West shipping routes.

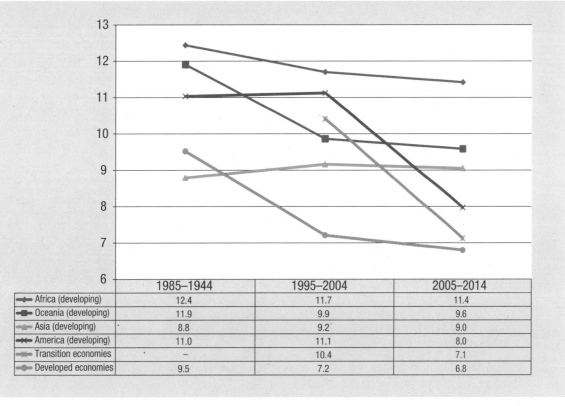

Figure 3.6. International transport costs: Freight costs as a percentage of value of imports, ten-year averages within country groups, 1985–2014

	1985–1944	1995–2004	2005–2014
Africa (developing)	12.4	11.7	11.4
Oceania (developing)	11.9	9.9	9.6
Asia (developing)	8.8	9.2	9.0
America (developing)	11.0	11.1	8.0
Transition economies	–	10.4	7.1
Developed economies	9.5	7.2	6.8

Source: UNCTAD secretariat estimates. Data represent the cost of international transport, excluding insurance costs, as a percentage of the "cost, insurance, freight" value of the imported goods.

Notes: Averages within the country groups are unweighted, that is, each country's freight ratio is assigned the same weight when calculating the average. Data are for all modes of transport.

(d) *Regulation and industry structure*: SIDS in Oceania, as well as several smaller African economies, only provide relatively small markets. As a result, shippers may be confronted with oligopolistic markets, where low levels of competition may lead to higher prices. In this context it would be a mistake to restrict competition by, for example, introducing any national or international cargo reservation regime.

(e) *Shipped product*: For manufactured goods, ad valorem transport costs tend to be lower than for raw materials, given the lower value per ton of raw materials compared to manufactured goods. As many developing countries from Africa and Oceania mostly import manufactured goods, the freight costs could actually be expected to be lower than in other regions. On the other hand, the types of manufactured goods imported into African and Oceanian developing countries tend to be of relatively lower value – that is, on average the cars, clothes or tools imported into Africa are of lower per unit value than those imported into Europe or North America. Hence, the cost of transport increases as a percentage share.

(f) *Port characteristics and infrastructure*: Many African and Oceanian developing countries, as well as those in Latin America and the Caribbean, are confronted with transport infrastructure bottlenecks. The largest ships that can be accommodated in most of these two regions' ports are far smaller than those that call at ports in other regions. Also, private sector participation through concessions is less frequent in developing countries such as those in Africa and Oceania. Both aspects contribute to higher transport costs. In this context, promoting regional/subregional hub ports that could handle larger vessels should be considered, including hub-and-spoke feedering, and interlining and relay services, as well as promoting private–public partnerships to upgrade and develop port infrastructure and cargo handling and operations.

(g) *Trade flows*: Most countries in Africa and Oceania have a merchandise trade deficit. As a consequence, ships are more likely to arrive fully loaded and have spare capacity when returning to China, Japan or Europe. Freight rates for imports should thus be higher than freight rates for exports. Given that figure 3.6 reflects data on imports, Africa and Oceania appear to have higher freight costs than the other regions. Although comprehensive data is not available, anecdotal evidence suggests that, effectively, freight rates for exports are lower than those for imports in most countries in these two regions.

In conclusion, the analysis of UNCTAD data on transport costs suggests that developing countries, especially in Africa and Oceania, pay more for the international transport of their imports than developed countries. The main reasons for this situation are to be found in these regions' trade imbalances, pending port and trade facilitation reforms, as well as lower trade volumes and shipping connectivity. There is potential for policymakers to partly remedy the situation through investments and reforms, especially in the regions' seaports, transit systems and customs administrations.

There is also a clear call for policymakers and port authorities to strengthen transnational cooperation to improve the development of the whole system, focusing on the causes that put a region or country at risk of becoming peripheral and uncompetitive. While there is not much that policymakers can do about their country's geographical position, some policy options exist to reduce costs by improving port infrastructure and increasing efficiency in the logistics chain, including through trade and transport facilitation, and more efficient port operations, or to become more attractive as a port of call, which would entail more port investments, and maritime transport service liberalization, as well as economic reforms to strengthen industry and trade relations.

C. RECENT DEVELOPMENTS IN FREIGHT RATES

In 2014, the freight rates market remained very volatile in its various segments. The continuous delivery of newly built large vessels and hesitant demand in the global shipping market put pressure on rates, as described below.

1. Container freight rates

Container freight rates remained volatile throughout 2014 although with different trends in individual trade lanes. Market fundamentals did not change significantly despite the expansion in global demand for container shipping (see chapter 1). This was mainly due to the constant supply pressures that the market rates continued to face with the introduction of very large units in mainlane trades and the cascading effect on non-mainlane trades (see chapter 2).

As shown in figure 3.7, the growth in global demand for container shipping reached 6 per cent in 2014 (compared to 5 per cent in 2013), outpacing that of supply, which remained at 5 per cent. Global container demand was boosted mainly by strong trade growth on the peak leg mainlanes of the Far East–Europe and the trans-Pacific, where North Europe imports and United States imports from Asia performed particularly well in 2014.

Mainlane freight rates recorded a general improvement in 2014 compared to 2013 levels. The Far East–Northern Europe rates averaged $1,161/TEU in 2014, up by 7 per cent from the 2013 average. In the trans-Pacific freight market, robust trade volumes as well as cargo diversions due to congestion problems at United States West Coast ports towards the end of 2014 improved freight rates on the Asia–United States East Coast lane. The Shanghai–United States East Coast freight rate averaged $3,720/40-foot equivalent unit (FEU) in 2014, 13 per cent higher than in 2013, compared to the Shanghai–West Coast route, which averaged $1,983/FEU, 3 per cent less than in 2013 (table 3.1).

Concerning non-mainlanes, freight rates performed less well as they also continued to face supply pressures from the cascade of large tonnage capacity coming from the mainlanes. On the North–South trades, where high levels of capacity redeployment have taken place, freight rates for Shanghai–South America averaged as low as $1,103/TEU in 2014, 20 per cent lower than in 2013. On the Shanghai–Singapore intra-Asian route, freight rates remained relatively flat, averaging around 1 per cent higher in 2014. The overcapacity also continued to influence the charter market and rates have remained more or less unchanged at low levels over the year.

In addition to cascading as a means of managing oversupply, carriers have continued to adopt idling and slow steaming (notwithstanding the decrease in fuel prices during the final months of 2014), which is estimated to have absorbed around 2.5 million TEUs of global nominal capacity. The idling of container ships fell to 0.2 million TEUs at the end of 2014, equivalent to 1.3 per cent of fleet capacity (Clarksons Research, 2015a).

At the same time, asset sales, consolidation and the cooperation efforts of several shipping lines to save on costs while improving efficiency and offering a worldwide network of routes have helped to improve operating margins in 2014. For instance,

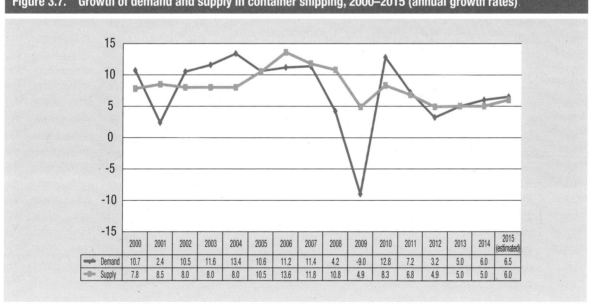

Figure 3.7. Growth of demand and supply in container shipping, 2000–2015 (annual growth rates)

	2000	2001	2002	2003	2004	2005	2006	2007	2008	2009	2010	2011	2012	2013	2014	2015 (estimated)
Demand	10.7	2.4	10.5	11.6	13.4	10.6	11.2	11.4	4.2	-9.0	12.8	7.2	3.2	5.0	6.0	6.5
Supply	7.8	8.5	8.0	8.0	8.0	10.5	13.6	11.8	10.8	4.9	8.3	6.8	4.9	5.0	5.0	6.0

Source: UNCTAD secretariat, based on data from Clarksons Research Container Intelligence Monthly, various issues.
Notes: Supply data refer to total capacity of the container-carrying fleet, including multipurpose and other vessels with some container-carrying capacity. Demand growth is based on million TEU lifts. The data for 2015 are projected figures.

Maersk Group, having launched a new sustainability strategy[1] in 2014 (which will run from 2014 to 2018), has seen its biggest contributor of overall emissions, Maersk Line, improve its efficiency by approximately 8 per cent in 2014 and save $98 million worth of fuel (Maersk Sustainability Report, 2014). Combined with reductions in unit costs (due to better vessel utilization and network efficiencies) and increases in volumes (while rates decreased by 1.6 per cent), Maersk Line reported a very satisfactory result of $2.3 billion profit in 2014 ($831 million higher than 2013).[2]

In addition, the plunge in fuel prices during the second half of 2014, including a steep fall in bunker fuel prices, with rates falling from $600 per ton in July 2014 to $250 in January 2015 (Barry Rogliano Salles, 2015) has also helped carriers boost their margins. In a survey covering 15 publicly traded carriers, it was noted that revenue decreased by 3 per cent compared with 2013, following a 5 per cent decline from 2012 (AlixPartners, 2015). In 2014, industry revenue remained more than 16 per cent below its 2008 peak of more than $200 billion.

Table 3.1. Container freight markets and rates

Freight markets	2009	2010	2011	2012	2013	2014
Trans-Pacific			($ per FEU)*			
Shanghai–United States West Coast	1 372	2 308	1 667	2 287	2033	1970
Percentage change		68.21	-27.77	37.19	-11.11	-3.10
Shanghai– United States East Coast	2 367	3 499	3 008	3 416	3290	3720
Percentage change		47.84	-14.03	13.56	-3.7	13.07
Far East–Europe			($ per TEU)			
Shanghai–Northern Europe	1 395	1 789	881	1 353	1084	1161
Percentage change		28.24	-50.75	53.58	-19.88	7.10
Shanghai–Mediterranean	1 397	1 739	973	1 336	1151	1253
Percentage change		24.49	-44.05	37.31	-13.85	8.86
North–South			($ per TEU)			
Shanghai–South America (Santos)	2 429	2 236	1 483	1 771	1380	1103
Percentage change		-7.95	-33.68	19.42	-22.08	-20.07
Shanghai–Australia/New Zealand (Melbourne)	1 500	1 189	772	925	818	678
Percentage change		-20.73	-35.07	19.82	-11.57	-17.11
Shanghai–West Africa (Lagos)	2 247	2 305	1 908	2 092	1927	1838
Percentage change		2.56	-17.22	9.64	-7.89	-4.62
Shanghai–South Africa (Durban)	1 495	1 481	991	1 047	805	760
Percentage change		-0.96	-33.09	5.65	-23.11	-5.59
Intra-Asian			($ per TEU)			
Shanghai–South-East Asia (Singapore)		318	210	256	231	233
Percentage change			-33.96	21.84	-9.72	0.87
Shanghai–East Japan		316	337	345	346	273
Percentage change			6.65	2.37	0.29	-21.10
Shanghai–Republic of Korea		193	198	183	197	187
Percentage change			2.59	-7.58	7.65	-5.08
Shanghai–Hong Kong (China)		116	155	131	85	65
Percentage change			33.62	-15.48	-35.11	-23.53
Shanghai–Persian Gulf (Dubai)	639	922	838	981	771	820
Percentage change		44.33	-9.11	17.06	-21.41	6.36

Source: Clarksons Research Container Intelligence Monthly, various issues.
Note: Data based on yearly averages.

The year 2014 also witnessed a reshaping of alliances. The failure of the P3 network between the three largest shipping companies, Maersk Line, Mediterranean Shipping Company and CMA CGM led to the creation of two important alliances: the 2M network, a 10-year vessel-sharing agreement between Maersk and the Mediterranean Shipping Company on the Asia–Europe and transatlantic routes; and the Three Ocean Alliance sharing agreement between CMA CGM, China Shipping Container Lines and United Arab Shipping Company, in a bid to save costs on key container routes between Asia and Europe, as well as Asia and North America. These alliances are expected to shift the industry towards the use of larger, more eco-efficient ships, particularly on Asia–Europe routes, and to yield cost savings by deploying larger and more efficient vessels and better utilization, coupled with lower CO_2 emissions.

Another important alliance is the global cooperation agreement between United Arab Shipping Company and Hamburg Süd that will give the Dubai-headquartered carrier access to South American trades, namely the Europe–South America east coast and Asia–South America east coast trades, and the German shipping line access to Asia–Europe and trans-Pacific trade, namely the Asia–North Europe and Asia–United States trades (*Lloyd's List Containerisation International*, 2014). Moreover, the German Hapag-Lloyd and the Chilean CSAV completed their merger, becoming the fourth-largest liner shipping company in the world.

For 2015, the container market can expect another challenging year. The order book schedule indicates that further ultra-large container ships will be delivered to the mainlanes in 2015–2016, and the extent to which cascading continues will largely determine freight rates on both the mainlane and non-mainlane trades. Moreover, some new challenges could emerge in the future, as global trade is expected to be increasingly concentrated around regional manufacturing hubs, thereby potentially decreasing future travel distances (Danish Ship Finance, 2015). The charter market environment may improve with significant scrapping levels of small and medium-sized vessels and the relatively small order book of container ship capacity in the smaller size ranges.

2. Tanker freight rates

The tanker market, which encompasses the transportation of crude oil, refined petroleum products and chemicals, witnessed an equally volatile freight rate environment in 2014. As a whole, the Baltic index for crude oil (Baltic Dirty Tanker Index) progressed by 21 per cent in 2014, reaching 777 points, whereas the Baltic Clean Tanker Index remained almost at the same level as in 2013, with 607 points, compared to 605 in 2013. In 2014, freight rates for both crude and product carriers increased in general for all vessel segments. Demand outperformed supply for the first time since 2010, leading to higher freight rates.

The crude tanker market turned out to be better than expected in 2014, particularly towards the second half of the year, when a drop in crude oil prices increased demand for such tankers. In addition, the slow expansion in oil fleet supply (which only increased by 4.5 per cent), slow steaming and the change in trading pattern (fewer imports to the United Sates and increasing demand from the Far East economies), which resulted in longer distances (Barry Rogliano Salles, 2015), triggered a surge in 2014 spot rates in most segments (tables 3.2 and 3.3).

The collapse in oil prices by almost 60 per cent over the second half of 2014 resulted in positive impacts on the tanker market. Demand for crude oil tankers was also boosted as a consequence of the increase in oil stockpiling, especially by Asian countries (namely China), increases in refinery runs and increases in floating storage as the contango situation developed.

Table 3.2.	Baltic Exchange tanker indices								
	2008	2009	2010	2011	2012	2013	2014	percentage change (2014/2013)	2015 (first half)
Dirty Tanker Index	1 510	581	896	782	719	642	777	21	853
Clean Tanker Index	1 155	485	732	720	641	605	607	0.33	678

Source: Clarksons Research, Shipping Intelligence Network – Timeseries, 2015.

Notes: The Baltic Dirty Tanker Index is an index of charter rates for crude oil tankers on selected routes published by the Baltic Exchange in London. The Baltic Clean Tanker Index is an index of charter rates for product tankers on selected routes published by the Baltic Exchange in London. Dirty tankers typically carry heavier oils, such as heavy fuel oils or crude oil. Clean tankers typically carry refined petroleum products such as gasoline, kerosene or jet fuels, or chemicals.

Table 3.3. Tanker market summary – clean and dirty spot rates, 2010–2014 (Worldscale)

Vessel type	Routes	2010 Dec.	2011 Dec.	2012 Dec.	2013 Dec.	2014 Jan.	Feb.	Mar.	Apr.	May	Jun.	Jul.	Aug.	Sept.	Oct.	Nov.	Dec.	Percentage change Dec. 2014/Dec. 2013
VLCC/ULCC (200 000 dwt+)	Persian Gulf–Japan	61	59	48	64	63	49	40	41	34	41	50	52	40	45	57	77	20.3%
	Persian Gulf–Republic of Korea	56	56	46	61	46	48	40	38	34	40	45	44	36	46	53	62	1.6%
	Persian Gulf–Caribbean/East Coast of North America	36	37	28	37	31	33	29	26	25	26	27	25	20	24	30	34	-8.1%
	Persian Gulf–Europe	57	59	26	..	n.a.	30	30	30	27	41	28	29	26	25	32	32	n.a.
	West Africa–China	..	58	47	61	57	54	45	42	39	40	48	51	45	49	59	63	3.3%
Suezmax (100 000–160 000 dwt)	West Africa–North-West Europe	118	86	70	102	109	59	62	60	58	70	85	69	59	76	102	91	-10.8%
	West Africa–Caribbean/East Coast of North America	103	83	65	97	102	57	60	60	52	64	81	63	56	79	91	79	-18.6%
	Mediterranean–Mediterranean	113	86	67	99	157	67	67	65	67	73	98	77	65	84	106	95	-4.0%
Aframax (70 000–100 000 dwt)	North-West Europe–North-West Europe	162	122	93	135	165	118	92	93	96	102	122	115	93	100	113	113	-16.3%
	North-West Europe–Caribbean/East Coast of North America	120	..	80	..	121	87	85	n.a.	70	80	90	n.a.	90	89	104	110	n.a.
	Caribbean–Caribbean/East Coast of North America	146	112	91	155	243	113	101	98	113	104	157	104	84	123	151	108	-30.3%
	Mediterranean–Mediterranean	138	130	85	100	167	87	94	92	81	81	98	100	85	92	166	106	6.0%
	Mediterranean–North-West Europe	133	118	80	107	204	83	89	87	65	74	98	104	79	92	185	108	0.9%
	Indonesia–Far East	111	104	90	99	109	97	86	86	87	96	101	110	93	90	105	116	17.2%
Panamax (40 000 - 70 000 dwt)	Mediterranean–Mediterranean	168	153	168	113	213	189	n.a.	118	n.a	n.a.	n.a.	128	120	100	n.a.	n.a.	n.a.
	Mediterranean–Caribbean/East Coast of North America	146	121	160	105	150	115	114	115	n.a.	n.a.	100	113	118	110	123	130	23.8%
	Caribbean–East Coast of North America/Gulf of Mexico	200	133	156	141	229	162	n.a.	109	121	114	162	147	118	113	148	150	6.4%
All clean tankers																		
70 000–80 000 dwt	Persian Gulf–Japan	125	105	116	81	73	78	88	90	91	82	87	116	108	114	115	102	25.9%
50 000–60 000 dwt	Persian Gulf–Japan	128	119	144	93	88	98	110	93	111	110	105	120	125	132	127	110	18.3%
35 000–50 000 dwt	Caribbean–East Coast of North America/Gulf of Mexico	158	155	162	..	103	105	101	100	96	91	142	100	95	100	131	175	n.a.
25 000–35 000 dwt	Singapore–East Asia	193	..	220	167	158	n.a.	168	180	n.a.	n.a.	n.a.	176	130	180	158	180	7.8%

Source: UNCTAD secretariat, based on Drewry Shipping Insight, various issues.
Notes: The figures are indexed per ton voyage charter rates for a 75,000 dwt tanker. The basis is the Worldscale 100.
* LCC: very large crude carrier; ULCC: ultralarge crude carrier.

CHAPTER 3: FREIGHT RATES AND MARITIME TRANSPORT COSTS

As such, the tight availability of tonnage and increase in activity pushed up very large crude carrier spot freight rates on key freight routes, namely Asian routes, towards the end of 2014. The spike in very large crude carrier earnings which began at the end of 2013 continued into 2014, hitting the highest levels since 2010. Very large crude carrier average spot earnings stood at $43,948/day for the last quarter of 2014 and $27,315/day for the entire year in 2014, an increase of 68 per cent from 2013. The Suezmax segment showed some significant movement, particularly in the last quarter of 2014, with growing West Africa–Europe trading being substituted for its primary West Africa–United States trade route – which was virtually eliminated. Supported by low oil prices, average yearly earnings rose by 79 per cent, reaching $27,791/day in 2014 (Clarksons Research, 2015b). Aframaxes benefited from tight tonnage and active trading in the Mediterranean–Caribbean/East Coast of North America and the Caribbean/East Coast of North America/Gulf of Mexico region (Drewry, 2015). Spot earnings averaged $24,705/day in 2014, a 75 per cent increase from the previous year.

For product tankers, while market rates improved towards the end of 2014 (mainly due to low crude oil prices that prompted demand for refinery runs, particularly in the United States and Asia–Pacific), they remained generally low during 2014. This was a result of imbalance between supply and demand that persisted in 2014, where supply growth (3.9 per cent) outpaced that of demand (2 per cent) in 2014. Nevertheless, clean spot yearly earnings declined by 2 per cent, reaching $12,361/day in 2014 (Clarksons Research, 2015b).

The tanker market is likely to remain positive in 2015, with low crude oil prices and increased storage trades. Nonetheless, the market is still suffering from overcapacity and freight rates will remain under pressure. Moreover, a change in the pattern of trade and demand, namely involving the decline in refining capacity in Europe and an increase in Asia and the Middle East, may result in increasing freight rate volatility. The Middle East has begun shifting its focus from crude oil exports to downstream developments such as refineries, making Atlantic basin crude oil (namely, South American suppliers) more attractive to Asia (Danish Ship Finance, 2015).

3. Dry bulk freight rates

Despite a strong start and high expectations for a positive impetus carried over from 2013, the dry bulk market freight rates faced another challenging year influenced by the surplus capacity that still exists and the uncertainties in demand projections in 2014. Bulk carrier earnings fell 5 per cent from 2013 to reach

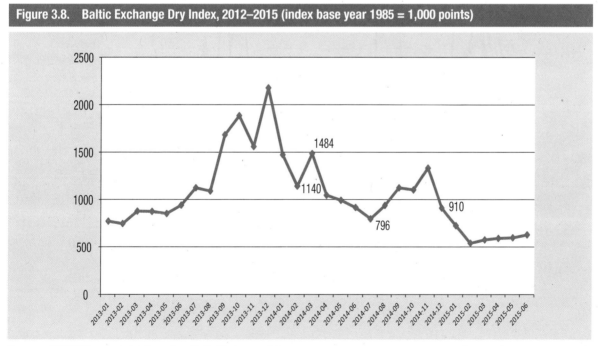

Figure 3.8. Baltic Exchange Dry Index, 2012–2015 (index base year 1985 = 1,000 points)

Source: UNCTAD secretariat, based on Baltic Exchange data.
Note: The index is made up of 20 key dry bulk routes measured on a time charter basis, and covers Handysize, Supramax, Panamax and Capesize dry bulk carriers, carrying commodities such as coal, iron ore and grain.

an average of $9,881/day. The low level of earnings exerted financial pressure on owners and led to several companies filing for bankruptcy (Clarksons Research, 2015b). As an overall indicator of the continued depression in dry bulk earnings, the Baltic Exchange Dry Index slid to a low level of 796 points in July 2014, to end at 910 points in December 2014 (figure 3.8).

Average earnings in the bulk carrier sector remained low and highly volatile in 2014 (figure 3.9). Capesize earnings during 2014 averaged $13,309/day, down 15 per cent from 2013. This was despite much faster growth in iron ore trade (driven by an increase in Chinese imports) than in the Capesize fleet (which grew by 4 per cent in 2014). The Panamax market continued to be negatively affected by oversupply of tonnage (driven by strong deliveries of Kamsarmaxes) and slower growth in coal trade (due to declining coal imports into China), with average earnings dropping down 5 per cent to $6,260/day and reaching as low as $2,137/day in June 2014. Supramax average earnings fell by 12 per cent to reach $10,819/day in 2014, dropping as low as $5,905/day in August before recovering for the remaining months and ending at $8,769/day (Clarksons Research, 2015c). The Indonesian ban on exports of unprocessed bauxite and nickel ore resulted in a weak Supramax market in the Far East.

The dry bulk market rates for 2015 and beyond will continue to be dominated by growing supply and uncertainties concerning the demand for dry bulk commodities from China. Factors that could influence demand in the future include innovation in technologies that seek to improve fuel efficiency and substitute for coal, and the increased number of countries that are setting policies and regulations aimed at reducing carbon emissions.

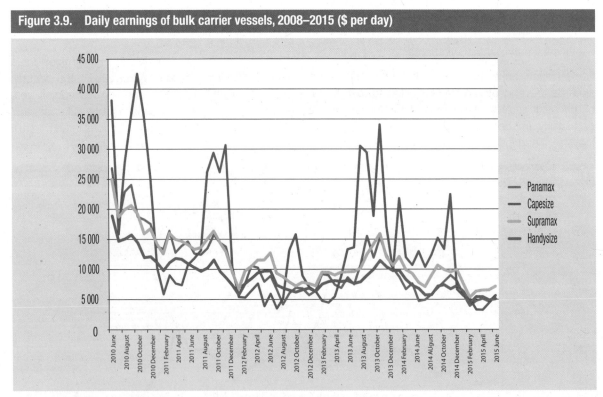

Figure 3.9. Daily earnings of bulk carrier vessels, 2008–2015 ($ per day)

Source: UNCTAD secretariat, based on data from Clarksons Research *Shipping Intelligence Network* and figures published by the London Baltic Exchange.

Note: Handysize – average of the six time charter routes; Supramax – average of the six time charter routes; Panamax – average of the four time charter routes; Capesize – average of the four time charter routes.

REFERENCES

AlixPartners (2015). *Container Shipping Outlook*. Available at http://www.alixpartners.com/en/LinkClick.aspx?fileticket=WD5LcjeJkhs%3d&tabid=635 (accessed 9 September 2015).

Angeloudis P, Bichou K, Bell M and Fisk D (2006). Security and reliability of the liner container shipping network: Analysis of robustness using a complex network framework. Presented at the International Association of Maritime Economists conference. Melbourne. 12–14 July.

Clarksons Research (2015a). *Container Intelligence Quarterly*. First quarter.

Clarksons Research (2015b). *Shipping Review and Outlook.* Spring.

Clarksons Research (2015c). *Dry Bulk Trade Outlook*. 21(1).

Collier P (2008). *The Bottom billion: Why the Poorest Countries are Failing and What Can Be Done About It*. Oxford University Press. Oxford.

Danish Ship Finance (2015). *Shipping Market Review*. May. Available at http://www.shipfinance.dk/en/SHIPPING-RESEARCH/~/media/PUBLIKATIONER/Shipping-Market-Review/Shipping-Market-Review---May-2015.ashx (accessed 18 September 2015).

Dicken P and Lloyd P (1998). *Standort und Raum – Theoretische Perspektiven in der Wirtschaftsgeographie*. Eugen Ulmer. Stuttgart, Germany: 95–123.

Drewry (2015). Analysis of the shipping markets. *Shipping Insight Monthly*. January.

ECLAC (1998). Concentration in liner shipping: Its causes and impacts for ports and shipping services in developing regions. LC/G.2027. Santiago. Available at http://www.cepal.org/en/publications/31094-concentration-liner-shipping-its-causes-and-impacts-ports-and-shipping-services (accessed 20 June 2015).

ECLAC (2002). The cost of international transport, and integration and competitiveness in Latin America and the Caribbean. *FAL Bulletin*. 191. Santiago. Available at http://repositorio.cepal.org/handle/11362/36199?show=full (accessed 20 June 2015).

Economic and Social Commission for Asia and the Pacific (2015). Reducing trade costs in Asia and the Pacific: Implications from the ESCAP–World Bank Trade Cost Database. Bangkok, 2015. Available at http://www.unescap.org/resources/reducing-trade-costs-implications-escap-world-bank-trade-cost-database (accessed 20 June 2015).

Hummels D and Schaur G (2013). Time as a trade barrier. *American Economic Review*. 103(7):2935–2959.

Kumar S and Hoffmann J (2002). Globalization: The maritime nexus. In: CT Grammenos, ed., *Handbook of Maritime Economics and Business*. Informa. Lloyds List Press. London.

Limao N and Venables A (2001). Infrastructure, geographical disadvantage, transport costs and trade. The World Bank Economic Review. 15(3):451–479.

Linnemann H (1966). *An Econometric Study of International Trade Flows*. North-Holland Publishing. Amsterdam.

Lloyd's List Containerisation International (2014). UASC and Hamburg Süd agree global co-operation. September. Available at http://www.lloydslist.com/ll/sector/containers/article449233.ece (accessed 9 September 2015).

Maersk Sustainability Report (2014). Available at http://www.maersk.com/~/media/annual-magazine-pj/maersk_sustainability_report_2014_online_version.pdf (accessed 9 September 2015).

Márquez-Ramos L, Martínez-Zarzoso I, Pérez-Garcia E and Wilmsmeier G (2005). Determinants of Maritime Transport Costs. Importance of Connectivity Measures. Presented at the International Trade and Logistics, Corporate Strategies and the Global Economy Congress. Le Havre. 28–29 September.

Martínez-Zarzoso I and Suárez Burguet C (2005). Transport costs and trade: Empirical evidence for Latin American imports from the European Union. *Journal of International Trade and Economic Development*. 14(3):227–45.

McCalla R, Slack B and Comtois C (2005). The Caribbean basin: Adjusting to global trends in containerization. *Maritime Policy and Management*. 32(3):245–261.

Palander T (1935). *Beiträge zur Standorttheorie*. Almqvist & Wiksell. Uppsala, Sweden.

Pöyhönen P (1963). A tentative model for the volume of trade between countries. *Weltwirtschaftliches Archiv*. 90:93–99.

Radelet S and Sachs J (1998). Shipping costs, manufactured exports, and economic growth. Paper presented at the American Economic Association Meeting, Harvard University. 1 January.

Sourdin P and Pomfret R (2012). *Trade Facilitation: Defining, Measuring, Explaining and Reducing the Cost of International Trade*. Edward Elgar Publishing. Cheltenham, United Kingdom.

Tinbergen J (1962). *Shaping the World Economy: Suggestions for an International Economic Policy*. Twentieth Century Fund. New York, United States.

UNCTAD (2015). The intrinsic relation between logistics performance and trade facilitation measures. Transport and Trade Facilitation Newsletter. First quarter. Issue No. 65. Available at http://unctad.org/en/PublicationsLibrary/webdtltlb2015d1_en.pdf (accessed 10 September 2015).

Wilmsmeier G (2003). Modal choice in South American freight transport: Analysis of constraint variables and a perspective for diversified modal participation in South America. Unpublished master's thesis. Technische Universität. Dresden, Germany.

Wilmsmeier G (2014). *International Maritime Transport Costs: Market Structures and Network Configurations*. Ashgate. Farnham, United Kingdom.

Wilmsmeier G and Hoffmann J (2008). Liner shipping connectivity and port infrastructure as determinants of freight rates in the Caribbean. *Maritime Economics and Logistics*. 10(1):130–151.

Wilmsmeier G, Hoffmann J and Sánchez RJ (2006). The impact of port characteristics on international maritime transport costs. In: Cullinane K and Talley W, eds. *Research in Transportation Economics*. Volume 16: Port Economics. Elsevier. Amsterdam.

Wilmsmeier G and Martínez-Zarzoso I (2010). Determinants of maritime transport costs – a panel data analysis. *Transportation Planning and Technology*. 33(1):117–136.

Wilmsmeier G and Sánchez RJ (2009). The relevance of international transport costs on food prices: Endogenous and exogenous effects. *Research in Transportation Economics*. 25(1):56–66.

ENDNOTES

[1] The Maersk sustainability strategy has three focus areas: enabling trade, energy efficiency and investing in education. See http://www.maersk.com/en/the-maersk-group/sustainability (accessed 9 September 2015).

[2] See Maersk line website news articles, available at http://www.maerskline.com/ur-pk/countries/int/news/news-articles/2015/02/maerskline-report-2014 (accessed 9 September 2015).

4 PORTS

Developing economies' share of world container port throughput increased marginally to approximately 71.9 per cent. This continues the trend of a gradual rise in developing countries' share of world container throughput. The increased share of world container throughput for developing countries reflects an increase in South–South trade.

The performance of ports and terminals is important because it affects a country's trade competitiveness. There are many determinants to port/terminal performance – labour relations, number and type of cargo handling equipment, quality of backhaul area, port access channel, land-side access and customs efficiency, as well as potential concessions to international terminal operators. The world's largest terminal operator handled 65.4 million 20-foot equivalent units (TEUs) in 2014, an increase of 5.5 per cent over the previous year. Of this figure, 33.6 million TEUs related to its operations at the port of Singapore and 31.9 million TEUs from its international portfolio. Hutchison Port Holdings trust is the second largest international terminal operator by market share. With operations in China and Hong Kong, China, it is not as geographically diverse as some other international terminal operators. APM Terminals has a geographical presence in 39 countries. DP World is the most geographically diverse of the global terminal operators, with a network of more than 65 terminals spanning six continents.

The economic, environmental and social challenges facing ports include growing and concentrated traffic volumes brought about by ever-increasing ship size; the cost of adaptation of port and port hinterland infrastructure measures; a changing marketplace as a result of increased alliances between shipping lines; national budget constraints limiting the possibilities of public funding for transport infrastructure; volatility in energy prices, the new energy landscape and the transition to alternative fuels; the entry into force of stricter sulphur limits (in, for example, International Maritime Organization (IMO) emission control area (ECA) countries); increasing societal and environmental pressure; and potential changes in shipping routes from new or enlarged international passage ways.

A. PORTS AND PORT-RELATED DEVELOPMENTS

Globally, there are a number of major developments under way that will have a direct impact on shipping and ports. For instance, construction of a second Suez Canal alongside the existing Suez Canal began in 2014 and continued into 2015. Traffic through the canal is expected to increase from an average of 49 ships per day to 97. Both transit times and waiting times will be reduced. For example, transit times will be shortened from 11 to 18 hours for the southbound convoy and the waiting time for vessels reduced from the present 8–11 hours to 3 hours. The New Suez Canal project is part of a major fiscal stimulus package meant to regain pre-2011 economic growth rates of around 7 per cent per year.

The development programme includes the creation of an industrial hub in adjacent areas, the development of five new seaports, a technology valley, and a centre for supplies and logistics. The project will cost an estimated $8.4 billion and is expected to more than double the canal's current annual revenue of $5 billion to $13 billion by 2023. Financing for the project was opened to Egyptian nationals, with a rate of return guaranteed at 12 per cent. The impact of the expansion of the Suez Canal on ports in the region is also likely to include an increase in the number of ships calling at the ports.

In contrast, the Panama Canal expansion project (see previous editions of the *Review of Maritime Transport*) is likely to be a game changer for regional ports as its expansion will allow for bigger vessels to transit. Bigger vessels mean more cargo, which means more revenue, but also increased adaption costs. Elsewhere, construction on the Nicaragua Canal has reportedly been delayed. An estimated $50 billion is needed to complete the construction (Gracie, 2015). A proposal to develop a canal through the isthmus of Thailand (Kra Canal) is also currently seeing another revival, having first been postulated 350 years ago. However, the proposal has not been officially confirmed (*Channel News Asia*, 2015). The cost of building the canal is estimated at $28 billion and, while it is technically feasible, the economic benefits have always remained uncertain as the time saving – an estimated three days (depending on speed) – is not as significant as 10 days for the Panama Canal and 20 days for the Suez Canal. In an era of economic uncertainty, vessel oversupply and the industry's response to slow-steam vessels, time saving is not the priority it once was. The cost to the environment and possible social tensions that may arise with any physical splitting of a country provide many reasons for careful analysis beyond mere economics.

1. Container ports

Container port throughput is measured by the number of TEUs that are handled. One FEU represents two TEU moves and the repositioning of containers to reach those stacked underneath/on top of others can also constitute a move. In chapter 1 it was observed that the number of full containers transported globally by sea in 2014 was estimated at 182 million, and yet the estimated port throughput is more than two and a half times that number, signifying that a lot of repositioning of empty containers occurs. The volumes reported in this chapter mainly relate to containerized cargo, which in turn represents more than half the value of all international seaborne trade and around one sixth of its volume. Container ports are multiple-user ports, that is, no one cargo owner has a monopoly of trade. Shipping lines may have dedicated terminals at which only they can call, but the cargo still has multiple owners. Other ports/terminals, for example for dry bulk and liquids, tend to be owned/operated by a single company that also owns the cargo. This is particularly so with commodity trade, where a large conglomerate may own an extraction mine, the railway, a processing plant and port facilities. The consequence of this is that operational data on bulk ports tend to be confidential and more difficult to ascertain. In addition, information on the volume and origin/destination of a particular commodity can affect its price in global markets as traders anticipate supply/demand levels, and thus industry practice tends to be selective in the information it releases to the public domain. Hence, this chapter mainly deals with container trade.

Chinese ports operate the largest number of berths (31,705) and handle more cargo both in terms of metric ton volume and number of TEUs than any other country. China's combined navigable rivers, at 126,300 kilometres, are also the longest of any single country. Understanding events in Chinese ports is thus a good indication of the global port industry. In 2014, Chinese river and sea ports handled 12.45 billion tons of cargo, an increase of 5.8 per cent over the previous year. Similarly, containerized cargo grew to 202 million TEUs, an increase of 6.4 per cent. China's major ports handled 2.7 billion tons of cargo, a modest increase of 2.2 per cent over the previous year. This slowdown in bulk imports is mostly driven by a weaker demand for major commodities, such as coal and iron ore (Yu, 2015).

CHAPTER 4: PORT DEVELOPMENTS

Table 4.1. Container port throughput for 80 developing countries/territories and economies in transition, 2012–2014 (TEUs)

Country/territory	2012	2013	Preliminary figures for 2014	Percentage change 2013–2012	Percentage change 2014–2013
China	161 318 524	170 858 775	181 635 245	5.91	6.31
Singapore	32 498 652	33 516 343	34 832 376	3.13	3.93
Republic of Korea	21 609 746	22 588 400	23 796 846	4.53	5.35
Malaysia	20 873 479	21 168 981	22 718 784	1.42	7.32
Hong Kong (China)	23 117 000	22 352 000	22 300 000	-3.31	-0.23
United Arab Emirates	18 120 915	19 336 427	20 900 567	6.71	8.09
Taiwan Province of China	14 976 356	15 353 404	16 430 542	2.52	7.02
Indonesia	9 638 607	11 273 450	11 900 763	16.96	5.56
India	10 279 265	10 883 343	11 655 635	*5.88*	7.10
Brazil	9 322 769	10 176 613	10 678 564	9.16	4.93
Viet Nam	7 509 119	9 036 095	9 424 699	20.33	4.30
Egypt	8 140 950	8 248 115	8 810 990	1.32	6.82
Thailand	7 468 900	7 702 476	8 283 756	3.13	*7.55*
Panama	7 217 794	7 447 695	7 942 291	3.19	6.64
Turkey	6 736 347	7 284 207	7 622 559	8.13	4.65
Saudi Arabia	6 563 844	6 742 697	6 326 861	2.72	-6.17
Philippines	5 686 179	5 860 226	5 869 427	3.06	0.16
Mexico	4 799 368	4 900 268	5 273 945	2.10	7.63
Islamic Republic of Iran	5 111 318	4 924 638	5 163 843	-3.65	4.86
Sri Lanka	4 321 000	4 306 200	4 907 900	-0.34	13.97
South Africa	4 360 100	4 694 500	4 831 462	7.67	2.92
Russian Federation	3 930 515	3 968 186	3 903 250	0.96	-1.64
Chile	3 596 917	3 722 980	3 742 520	3.50	0.52
Oman	4 167 044	3 930 261	3 620 364	*-5.68*	*-7.88*
Colombia	2 991 941	2 746 038	3 127 994	-8.22	13.91
Morocco	1 826 100	2 558 400	3 070 000	40.10	*20.00*
Pakistan	2 375 158	2 485 086	2 597 395	4.63	*4.52*
Peru	2 031 134	2 086 335	2 234 582	*2.72*	*7.11*
Costa Rica	1 329 679	1 880 513	1 960 267	41.43	*4.24*
Dominican Republic	1 583 047	1 708 108	1 795 221	7.90	*5.10*
Ecuador	1 594 711	1 675 446	1 786 981	5.06	6.66
Argentina	1 986 480	2 141 388	1 775 574	7.80	*-17.08*
Bangladesh	1 435 599	1 500 161	1 655 365	4.50	10.35
Jamaica	1 855 400	1 703 900	1 638 100	*-8.17*	*-3.86*
Bolivarian Republic of Venezuela	1 249 500	1 348 211	1 416 970	*7.90*	*5.10*
Bahamas	1 202 000	1 400 000	1 399 300	16.47	-0.05
Kuwait	1 126 668	1 215 675	1 277 674	7.90	*5.10*
Guatemala	1 158 400	1 211 600	1 273 392	*4.59*	*5.10*
Lebanon	882 922	1 117 300	1 210 400	26.55	8.33
Nigeria	877 679	1 010 836	1 062 389	15.17	5.10
Kenya	903 400	894 000	1 010 000	-1.04	12.98
Angola	750 000	913 000	1 000 000	21.73	9.53
Uruguay	753 000	861 000	904 911	*14.34*	*5.10*
Yemen	760 192	820 247	862 079	*7.90*	*5.10*
Ukraine	748 889	808 051	849 262	*7.90*	*5.10*
Syrian Arab Republic	737 448	795 707	836 288	*7.90*	*5.10*

Table 4.1. Container port throughput for 80 developing countries/territories and economies in transition, 2012–2014 (TEUs) *(continued)*

Country/territory	2012	2013	Preliminary figures for 2014	Percentage change 2013–2012	Percentage change 2014–2013
Ghana	735 229	793 312	833 771	*7.90*	*5.10*
Jordan	703 354	758 919	797 624	*7.90*	*5.10*
Côte d'Ivoire	690 548	745 102	783 102	*7.90*	*5.10*
Djibouti	681 765	735 624	773 141	*7.90*	*5.10*
Trinidad and Tobago	651 332	702 787	738 630	7.90	5.10
Honduras	665 354	670 726	704 934	0.81	*5.10*
Mauritius	576 383	621 917	653 635	7.90	*5.10*
United Republic of Tanzania	487 786	526 321	638 023	*7.90*	21.22
Tunisia	529 956	571 823	600 986	*7.90*	*5.10*
Sudan	498 938	538 354	565 811	*7.90*	*5.10*
Libya	369 739	434 608	456 773	17.54	5.10
Senegal	396 822	428 171	450 008	*7.90*	*5.10*
Qatar	393 151	424 210	445 845	*7.90*	*5.10*
Congo	385 102	415 525	436 717	*7.90*	*5.10*
Benin	359 908	388 341	408 146	*7.90*	*5.10*
Papua New Guinea	337 118	363 750	382 301	*7.90*	*5.10*
Bahrain	329 470	355 498	373 628	*7.90*	*5.10*
Cameroon	323 917	349 507	367 332	*7.90*	*5.10*
Algeria	317 913	343 028	360 522	*7.90*	*5.10*
Mozambique	289 411	312 274	328 200	*7.90*	*5.10*
Cuba	265 281	286 238	300 836	*7.90*	*5.10*
Georgia	256 929	277 226	291 365	*7.90*	*5.10*
Cambodia	254 760	274 886	288 905	*7.90*	*5.10*
Myanmar	215 945	233 005	244 888	*7.90*	*5.10*
Guam	208 181	224 628	236 084	*7.90*	*5.10*
Gabon	174 597	188 390	197 998	*7.90*	*5.10*
El Salvador	161 000	180 600	189 811	12.17	*5.10*
Madagascar	160 320	172 986	181 808	*7.90*	*5.10*
Croatia	155 724	168 026	176 596	*7.90*	*5.10*
Aruba	147 716	159 385	167 514	*7.90*	*5.10*
Namibia	115 676	124 815	131 180	*7.90*	*5.10*
Brunei Darussalam	112 894	121 813	128 026	*7.90*	*5.10*
New Caledonia	102 423	110 514	116 150	*7.90*	*5.10*
Nicaragua	93 737	96 472	101 392	2.92	*5.10*
Subtotal	443 672 437	466 256 062	491 169 015	5.09	5.34
Other reported	689 351	739 276	761 420	7.24	3.00
Total reported	444 361 788	466 995 338	491 930 435	5.09	5.34
World Total	624 480 174	651 200 742	684 429 339	4.28	5.10

Source: UNCTAD secretariat, derived from various sources including Dynamar B.V. publications and information obtained by the UNCTAD secretariat directly from terminal and port authorities.

Notes: Singapore includes the port of Jurong. The term "other reported" refers to countries/economies with fewer than 100,000 TEUs per year. Many figures for 2013 and 2014 are UNCTAD estimates (these figures are indicated in italics). Country totals may conceal the fact that minor ports may not be included; therefore, in some cases, the actual figures may be different than those given.

CHAPTER 4: PORT DEVELOPMENTS

In the first quarter of 2015, Chinese ports handled 49 million TEUs, an increase of 7.3 per cent over the same period in the previous year. This was largely due to a recovery in the United States economy. The figures would suggest that the major Chinese exporting ports experienced a significant growth while the growth of importing ports (for example, in bulk cargo) has slowed. This could mean that factories are reducing their stockpiles in anticipation of a slow growth in the world economy.

Table 4.1 lists the container throughput of 80 developing countries and economies in transition with a national throughput greater than 100,000 TEUs (port throughput figures for 126 countries/territories are available at http://stats.unctad.org/TEU). In 2014, the container throughput for developing economies grew by an estimated 5.34 per cent to 491 million TEUs. This growth is higher than the 5.1 per cent seen in the previous year. The container throughput growth rate for all countries in 2014 is estimated at 684.4 million TEUs, a rise of 5.1 per cent over the previous year.

Developing economies' share of world throughput increased by 0.2 per cent to approximately 71.9 per cent. This continues the trend of a gradual rise in developing countries' share of world container throughput. The two main drivers of this process are developing countries' greater participation in global value chains and the continued increase of containers for transporting dry bulk cargo.

Table 4.2 shows the world's 20 leading container ports for the period 2012–2014. The top 20 container ports accounted for approximately 45.7 per cent of world container port throughput in 2014. These ports showed a 4.5 per cent increase in throughput compared to 2013, the same as the estimated increase for 2013. The list includes 16 ports from developing economies, all of which are in Asia; the remaining four ports are from developed countries, three of which are located in Europe and one in North America. All of the top 10 ports continue to be located in Asia, signifying the importance of the region as a manufacturing hub. Ningbo remained in fifth position but achieved the highest growth at 12 per cent, a growth rate closely followed by Dubai and Tanjung Pelepas. The port of Tanjung Pelepas moved up two places to eighteenth position following completion of infrastructure investments. The port of Long Beach was displaced from the top 20 list due to low growth as a result of labour disputes at the port and the higher

Table 4.2. Top 20 container terminals and their throughput, 2012–2014 (TEUs and percentage change)

Port Name	2012	2013	2014	Percentage change 2013–2012	Percentage change 2014–2013
Shanghai	32 529 000	36 617 000	35 290 000	12.57	-3.62
Singapore	31 649 400	32 600 000	33 869 000	3.00	3.89
Shenzhen	22 940 130	23 279 000	24 040 000	1.48	3.27
Hong Kong	23 117 000	22 352 000	22 200 000	-3.31	-0.68
Ningbo	15 670 000	17 351 000	19 450 000	10.73	12.10
Busan	17 046 177	17 686 000	18 678 000	3.75	5.61
Guangzhou	14 743 600	15 309 000	16 610 000	3.83	8.50
Qingdao	14 503 000	15 520 000	16 580 000	7.01	6.83
Dubai	13 270 000	13 641 000	15 200 000	2.80	11.43
Tianjin	12 300 000	13 000 000	14 060 000	5.69	8.15
Rotterdam	11 865 916	11 621 000	12 298 000	-2.06	5.83
Port Klang	10 001 495	10 350 000	10 946 000	3.48	5.76
Kaohsiung	9 781 221	9 938 000	10 593 000	1.60	6.59
Dalian	8 064 000	10 015 000	10 130 000	24.19	1.15
Hamburg	8 863 896	9 258 000	9 729 000	4.45	5.09
Antwerp	8 635 169	8 578 000	8 978 000	-0.66	4.66
Xiamen	7 201 700	8 008 000	8 572 000	11.20	7.04
Tanjung Pelepas	7 700 000	7 628 000	8 500 000	-0.94	11.43
Los Angeles	8 077 714	7 869 000	8 340 000	-2.58	5.99
Jakarta	6 100 000	6 171 000	6 053 000	1.16	-1.91
Total top 20	284 059 418	296 791 000	310 116 000	4.48	4.49

Source: UNCTAD secretariat, based on Dynamar B.V., June 2015, and various other sources.
Note: Singapore does not include the port of Jurong.

rates of growth of other ports. Jakarta port was a new entrant to the list as a result of a continued steady increase in demand that has seen throughput at the port grow by more than 50 per cent since 2009 due to the buoyant economy (*Drewry*, 2015).

B. INTERNATIONAL TERMINAL OPERATORS

1. Operational performance

The performance of ports and terminals can significantly affect a country's trade competitiveness. One chief economist even cited port congestion as the new barrier to international trade (van Marle, 2015). There are many determinants to port/terminal performance – for example, labour relations, number and type of cargo handling equipment, quality of backhaul area, port access channel, land-side access, customs efficiency, and the like. These specific operational indicators are generally more useful to port operators and do not include non-tangible assessments (for example, users' perceptions, service quality, innovation levels, and the like) that port customers may find more beneficial (Cetin, 2015).

Terminal operators rarely publish their performance ratings, but are sometimes obliged to do so due to publicity, for example Malaysia's Westports "set a new world record for container terminal productivity, notching an impressive 793 moves in one hour over the CSCL [China Shipping Container Lines] Le Havre (9,572 TEU vessel) with the deployment of nine twin-lift cranes" (Westports, 2015). Ports and terminals rarely publish data on their performance that allow shippers to make informed choices or policymakers to identify best practices. While there may be many reasons for this, such as no statutory requirement or limited readership, the strongest reason is likely to be the unnecessary scrutiny it would generate without any immediate return. In an age where many companies' chief executive officers have limited time in their positions and short reporting periods the situation is unlikely to change. However, international pressure, for instance in the area of sustainability reporting, may help to change this situation. Until then it tends to be the customers who report on the performances of their service providers. For instance, Drewry Shipping Consultants has launched its Drewry Benchmarking Club. The club is limited to importers and exporters (that is, buyers of shipping services) and excludes providers of shipping services (carriers) and intermediaries/brokers (forwarders/non-vessel operating common carriers). While it aims to benchmark ports and routes, its primary focus seems to be on freight costs. The *JOC* recently produced its port productivity rankings, which examine loading/unloading data from 17 carriers at over 500 ports worldwide. From these two initiatives it is clear that it is the ports' customers (that is, shippers and carriers) who are sharing information for their mutual benefit about the ports' performance. Ports may be forced to publish their own data should they not agree with how their customers are assessing them. Table 4.3 shows the ranking of port terminals in 2014, with Yokohama ranking as the world's most efficient container port, having increased productivity by 10 per cent over the previous year. Unlike other terminals, APM Terminals Yokohama has been successful in improving its efficiency year after year due to the synchronized process developed between the vessel and the container yard that eliminates virtually all wasted time between the quay crane and yard equipment operations.

Table 4.4 shows the productivity ranking of ports in 2014 and the change over the preceding two years. Some ports are home to several terminal operators, thus providing intra-port competition. For example, the port of Tianjin, which is ranked in second place, is home to numerous international terminal operators,

Table 4.3. Top global terminals' berth productivity, 2014 (container moves per ship, per hour on all vessel sizes)

Terminal	Port	Country	Berth productivity
APM Terminals Yokohama	Yokohama	Japan	180
Tianjin Port Pacific International	Tianjin	China	144
DP World-Jebel Ali Terminal	Jebel Ali	United Arab Emirates	138
Qingdao Qianwan	Qingdao	China	136
Tianjin Port Alliance International	Tianjin	China	132
Ningbo Beilun (second)	Ningbo	China	127
Guangzhou South China Oceangate	Nansha	China	122
Busan Newport Co. Ltd.	Busan	Republic of Korea	119
Yantian International	Yantian	China	117
Nansha Phase I	Nansha	China	117

Source: JOC Port Productivity Database 2015.

CHAPTER 4: PORT DEVELOPMENTS

such as APM Terminals, China Merchants Holdings International, COSCO Pacific, CSX World Terminals OCCL, PSA and DPW. Interestingly, while all the ports in this table experienced productivity gains of between 30 and 60 per cent in 2013 over the previous year, in 2014 only three ports managed to continue the upward improvement. This suggests that port performance and continued improvement are still difficult to achieve.

In a study involving 203 ports in 70 developing countries, with 1,750 data points, it was observed that operational changes rather than scale efficiency (the process of adding more equipment) resulted in increases in port efficiency. It should be noted that pure efficiency is the result of input divided by output. With regard to ports, inputs may be numerous and difficult to calculate (for example, utilized space, multiple currencies' operational hours and the like). Most port-related studies avoid this shortcoming by measuring productivity (output) over a certain period. Both efficiency and productivity tend to be referred to interchangeably to a large extent. From 2000 to 2010 there was an upward trend in increasing port efficiency within developing regions, from 47 per cent to 57 per cent. The main determinants were private sector participation, the reduction of corruption in the public sector and improvements in liner connectivity, as well as the increased provision of multimodal links that led to an increase in the level of port efficiency in developing regions (Suárez-Alemán et al., 2015). Port performance matters the most on a regional basis where there is a real possibility that cargo can move to a competing, more efficient port. A study of ports in West Africa showed that they exhibited high levels of efficiency and that four out of six ports had an average efficiency score of 76 per cent or higher for the period under study (van Dyck, 2015). Yet in another study by the *JOC* for all Africa, African ports were on average ranked as the least productive of all regional groups (*Data in Motion*, 2015). The poor performance of port management and operations, together with other procedural inefficiencies along the logistics chain, and imbalanced freight rates that shipping lines charge because of empty backhaul cargo, are all contributing factors to high transport costs (Bofinger et al., 2015). Every minute that a vessel stays at a terminal means money lost for the shipping company, and this in turn places pressure upon the terminal operator to ensure it does not lose business to more efficient competitors (ACS–AEC, 2015). Port privatization is often seen as the best means to bring in private sector expertise and turn around the performance of a port. Many countries privatized their ports in the 1990s, but there are still many State-owned and operated ports around the world. In Viet Nam, the Government plans to privatize an estimated 432 State-owned enterprises during the period 2014–2015, including 19 seaports (*Vietnam Briefing*, 2015).

When Governments review proposals for new port infrastructure projects it is difficult for them to judge whether the traffic volumes and marginal cost savings will match predictions. In a recent survey of around 500 terminals worldwide it was observed that the average TEU per metre of quay per year was 1,072, while the TEU per hectare was 24,791 and TEU per gantry crane 123,489 (Drewry, 2014b). Some of the worst performing ports per TEU, hectare and crane utilization were in North America. Varying levels of

Table 4.4. World's leading ports by productivity, 2014 (container moves per ship, per hour on all vessel sizes and percentage increase)

Port	Country	2012 berth productivity	2013 berth productivity	2014 berth productivity	Percentage increase 2013/2012	Percentage increase 2014/2013
Jebel Ali	United Arab Emirates	81	119	138	47%	16%
Tianjin	China	86	130	125	51%	-4%
Qingdao	China	96	126	125	31%	-1%
Nansha	China	73	104	119	42%	14%
Yantian	China	78	106	117	36%	10%
Khor al Fakkan	United Arab Emirates	74	119	108	61%	-9%
Ningbo	China	88	120	107	36%	-11%
Yokohama	Japan	85	108	105	27%	-3%
Busan	Republic of Korea	80	105	102	31%	-3%
Xiamen	China	76	106	90	39%	-15%

Source: UNCTAD secretariat and *JOC* Port Productivity Database 2015.

cargo volumes, trans-shipment share and automation of processes all contributed to the outcome. While the provision of more space or bigger cranes is not a guarantee for additional cargo, it is useful for policymakers to know when examining project proposals what they can expect from proposed new facilities. Interestingly, the study also shows that, on average, gantry crane productivity tends to be about 50 per cent of the maximum capacity advertised by the manufacturer. This could have a financial impact upon ports when planning future improvements.

According to one study, the largest liner shipping company, Maersk Line, makes around 31,000 port calls, with 1,500–1,800 moves per call, and spends some 19 per cent of its total costs on ship fuel. A 7 per cent reduction in port stay during a 13–18-hour call would allow the company to steam slower once a vessel leaves port and reduce fuel consumption by around $120 million per year (van Marle, 2015). The reduction in a ship's time in port primarily depends on the performance of the port in fulfilling its functions.

2. Financial performance

The traditional role of ports as gateways between foreign and domestic markets has meant that growth in throughput and revenue for a port is reliant upon external factors beyond the control of the port, such as the ability of the port's hinterland to either import or export more goods. For terminal operators, replicating home-grown efficiencies in foreign markets can be an ideal way for the businesses to expand when faced with domestic limitations beyond their control. Many terminal operators have expanded horizontally (for example, doing the same thing in a different place) or vertically (for example, by controlling different aspects of a supply chain). Presently there are numerous owners of terminal operators that control ports on a worldwide basis. Together, the leading global container terminals accounted for around 300 million TEUs in 2013, or around 47 per cent of the world's container port throughput (Drewry, 2014b).

The world's largest terminal operator, PSA International (formally the Port of Singapore Authority) handled 65.4 million TEUs in 2014, an increase of 5.5 per cent over the previous year. Of this figure, 33.6 million TEUs are accounted for by its operations in the port of Singapore (+4.2 per cent) and 31.9 million TEUs by its international portfolio (+7.2 per cent). Its international portfolio stretches across 16 countries and three continents. However, it does not operate terminals in Africa, Australia or North America. Revenue for the company grew slightly in 2014 to $3.8 billion, whereas profit slightly decreased to $1.4 billion (PSA, 2014). Among the major terminal operators, PSA International is the market leader in terms of not only market share of global port throughput, but also the ratio of revenue to profits.

Hutchison Port Holdings Trust is the second largest international terminal operator by market share. With operations in China, including Hong Kong (China), it is not as geographically diverse as some other international terminal operators. Its 2014 throughput of approximately 24.2 million TEUs was up 6.3 per cent over the previous year. Revenue increased 1.9 per cent to HK$12.6 billion ($1.63 billion) for 2014, while operating profit increased 5.5 per cent to HK$4.2 billion ($540 million).

APM Terminals has a geographical presence in 39 countries. This includes 65 port and terminal facilities and 200 inland services. In 2014, its revenue was the highest of all international terminal operators at $4.5 billion, an increase of 2.7 per cent, while internal efficiencies pushed operating profit to $900 million, an increase of 14.4 per cent from the previous year despite substantial losses in its Russian business. Of the leading global terminal operators, APM Terminals has seen the biggest impact of international sanctions placed on the Russian Federation. To illustrate this, volumes from Asia to Russian Black Sea ports dropped almost 50 per cent in the first four months of 2015, compared with the same period in 2014 (*Lloyd's List – Daily Briefing*, 2015). APM Terminals has a 30.75 per cent stake in Global Ports, the Russian Federation's leading operator, with seven maritime container terminals representing about half of the country's annual throughput. Financial shares in Global Ports dropped almost 80 per cent from $16 per share to just $3 in the year following the start of the crisis (Pasetti, 2015).

DP World is the most geographically diverse of the global terminal operators with a network of more than 65 terminals spanning six continents. Recent new projects include DP World London Gateway and Embraport (Brazil), which both became operational in 2013. Expansion to existing facilities occurred with the opening of terminal 3 at its home port of Jebel Ali in the United Arab Emirates and a new container terminal at Southampton in the United Kingdom. In 2014, it handled 60 million TEUs, an increase of 8.9 per cent over the previous year. In 2014, revenue increased by 10 per cent to $3.4 billion and profit by a similar growth rate to $675 million.

CHAPTER 4: PORT DEVELOPMENTS

From the above brief overview of the leading container terminal operators it can be seen that the enterprise is profitable. The top four global terminal operators combined generated $3.5 billion in profit in 2014 on total revenues of $13.3 billion, an average return of 26 per cent. For policymakers this poses a challenge – profits earned by international terminal operators increase transport costs, which can affect national competitiveness. Yet by having an efficient port and being better connected to international markets, transport costs could be lower than otherwise possible. Ideally, having inter-port competition between multiple ports is best, or where this is not possible, intra-port competition with the presence of multiple terminal operations in one port, could help keep transport costs low. Some countries such as India and South Africa have set limits on the tariffs terminal operators are allowed to charge, although these have met with mixed results. Another issue to consider is that global terminal operators must be financially empowered to address the increasing costs associated with meeting sustainable development criteria.

C. SUSTAINABILITY CHALLENGES FACING PORTS

The economic, environmental and social challenges facing ports include: growing and concentrated traffic volumes brought about by ever-increasing ship size; the cost of adaptation of port and port hinterland infrastructure measures; a changing marketplace as a result of increased alliances between shipping lines; national budget constraints limiting the possibilities of public funding for transport infrastructure; volatility in energy prices, the new energy landscape and the transition to alternative fuels; entry into force of the stricter sulphur limits in, for example, IMO ECA countries; increasing societal and environmental pressure; potential changes in shipping routes from enlarged or new international passages (for example, the existing Suez and Panama Canals, and new proposals such as the Nicaragua and Kra Canals mentioned earlier); an uncertain geopolitical situation and its effect on shifting supply chains; further globalization of business and society; and barriers to internal markets (for example, customs inspection) for maritime transport.

1. Environmental challenges

The transportation industry's share of all the global energy consumed increased from 45 per cent in 1973 to 62 per cent in 2010 (Hui-huang, 2015). In terms of emissions, it is second only to the energy consumed to regulate indoor temperature. In 1992, UNFCCC considered how countries could limit industrial emissions and the anticipated planetary climate change. However, it was realized that emission reduction provisions in the Convention were inadequate and, as a result, new measures strengthening the global response to climate change were adopted under the 1997 Kyoto Protocol. The Kyoto Protocol, which entered into force on 16 February 2005, legally binds developed countries to emission reduction targets. There are now 195 Parties to the Convention and 192 Parties to the Protocol. Parties to the Protocol have continued the negotiations and have amended it to achieve more ambitious results. The Kyoto Protocol in effect "operationalizes" UNFCCC by committing industrialized countries to stabilize GHG emissions. It operates on the principle of "common but differentiated responsibility", where certain countries are obliged to make emission reductions in recognition of their contribution to the existing GHGs. Emissions from national maritime transport (for example, inland waterways, lakes and coastal shipping) and port emissions are included in the Kyoto Protocol. Emissions of CO_2 by shipping were estimated at 3.3 per cent of global emissions for 2007 (IMO, 2015). Greenhouse gas emissions produced from international maritime transport – for example, while sailing in international waters – are, however, not included in the Kyoto Protocol. International maritime transport emissions are estimated at 83 per cent of all shipping emissions (Villalba and Gemechu, 2011). The Kyoto Protocol recognizes that, concerning maritime issues, countries must work through IMO. However, IMO works on the principle of "no less favourable treatment of ships", which means ships must not be placed at a disadvantage because their country has or has not ratified a convention. Thus, in the field of environmental protection, ports face a complicated regulatory requirement as well as societal expectations (Lam and Notteboom, 2014). Such pressure can have an impact on the further space for the ports to grow, not only in terms of hectares, but also in terms of the "environmental space" concept.[1] This means that tackling maritime-related emissions is complicated and that valuable time may be spent interpreting text (Fitzgerald et al., 2011).

The ports with the greatest absolute emissions attributable to shipping are Singapore, Hong Kong (China), Tianjin (China) and Port Klang (Malaysia). The distribution of shipping emissions in ports is skewed:

the 10 ports with the greatest emissions represent 19 per cent of total CO_2 emissions in ports and 22 per cent of SOx emissions. The port with the lowest relative CO_2 emissions (emissions per ship call) is Kitakyushu (Japan); the port of Kyllini (Greece) has the lowest SOx emissions. Other ports with relatively low emissions are situated in Greece, Japan, Sweden, the United Kingdom and the United States (Merk, 2014).

Shipping emissions in ports are substantial, accounting for 18 million tons of CO_2, 0.4 million tons of NOx, 0.2 million of SOx and 0.03 million tons of "PM10" (particulate matter with diameter inferior to 10 micrometres) in 2011. Around 85 per cent of ships' emissions are attributable to two ship types, container ships and tankers. It is estimated that most shipping emissions in ports (CH_4, CO, CO_2 and NOx) will grow fourfold until 2050. Asia and Africa are expected to see the sharpest increases in emissions, due to strong port traffic growth and limited mitigation measures (Merk, 2014).

On 1 January 2015, IMO regulation 14 of annex VI of MARPOL on ship emissions came into force. The regulation is intended to improve air quality by limiting the sulphur content of fuels used by ships operating in ECAs, including ports, to 0.10 per cent. This will require existing vessels to switch to lower sulphur content fuel while in an ECA or retrofit vessels with scrubbers to clean the exhaust fumes before they enter the atmosphere. Scrubbing uses a fluid containing alkaline material that absorbs SOx and neutralizes them. The remaining exhaust gases are then released and the residue waste sludge is stored on board until it can be transferred ashore and safely disposed of. New vessels are, however, being built to ensure that they are fully compliant with MARPOL regulations. While the impact of the new regulation is not yet clear, some transport service providers believe that its immediate effect will be to increase transport costs and move short-haul cargo from sea to road. Outside the ECAs, the sulphur content of fuels is capped at 3.5 per cent but set to be reduced to 0.50 per cent from 1 January 2020 (or 2025, depending on the enforcement date and the result of an IMO review on the availability of low sulphur fuels). European ports have much lower emissions of SOx (5 per cent) and PM10 (7 per cent) than their share of port calls (22 per cent), which may be explained by the European Union regulation to use low sulphur fuels at berth (Merk, 2014).

During their stay in ports, ships emit pollutants such as CO_2, SO_2, NO_X (the sum of NO and NO_2 emissions) and, in smaller amounts, CO, PM, non-CH_4 volatile organic compounds, CH_4 and N_2O (Fitzgerald et al., 2011). Other pollutants include dust from bulk cargo handling, emissions related to electricity consumption, and gases from cargo handling equipment and trucks (Economic and Social Commission for Asia and the Pacific, 1992; and Villalba and Gemechu, 2011). Vibration, light pollution and wake damage also give rise to a variety of issues. Ports tend to be seen as sources of pollution because they are easily identifiable, immovable and close to the community most affected by the effects of the pollution. Health effects include bronchitic symptoms that have been linked to NO_2 and CO emissions, while exposure to SO_2 is associated with respiratory issues and premature births (Merk, 2014). Ports need employees from the local community and employees need their jobs, thus their relationship is much closer than it is to ship operators. This means that cooperation between ports and their local communities is mutually beneficial and easier to facilitate. The challenge for ports is that communities have, through the advent of the Internet, become more empowered with access to scientific information, more vocal and better organized. A port authority thus needs to ensure not only that it provides a safe working environment that protects workers' health and promotes their personal development but also provides social responsibility, ethical governance and accountability. The port authority must show it manages environmental risks well and furthers the economic and social development of the surrounding region, as well as promotes equality and respect for cultural diversity through the involvement of stakeholders in port development and operations (Doerr, 2011). For ports, the usual three pillars of sustainability (economic, environmental and social) must be entwined with an institutional dimension to cater for multiple stakeholders.

The 2012 United Nations Conference on Sustainable Development, known as Rio+20, acknowledged in its outcome document (known as "The Future We Want") the importance of corporate sustainability reporting and encouraged companies, especially large or publicly listed companies, to consider integrating sustainability information into their reporting cycles. To this end, UNCTAD was designated as one of the official implementing bodies for action on sustainability reporting, primarily through its role as the host of the Intergovernmental Working Group of Experts on International Standards of Accounting and Reporting. In 2014, UNCTAD published a report, entitled *Best Practice Guidance for Policymakers and Stock Exchanges on Sustainability Reporting Initiatives*, recognizing the role stock exchanges have in influencing companies. This report cited as an example the fact that disclosure of "policy and performance in connection with environmental and social responsibility" was only mandatory in 56 per cent of 25 emerging markets, yet it was voluntarily reported by 91 per cent of 188 of the largest companies in those markets. Thus, mandatory rules are not necessarily the only course of action for policymakers – simply nudging businesses in a particular direction can be more effective.

Sustainability reporting for ports and terminals is still in its early stages. Key issues to address include the reduction of kilograms of CO_2 emitted per modified TEU ($kgCO_2e/modTEU$), the reduction in megajoules of energy used per total terminal moves, and the reduction in fresh water consumption for cleaning equipment. One terminal operator, DP World, reduced its fresh water consumption by 75 per cent (64 million litres) by installing water recycling plants. DP World's sustainability reporting also announced that the intensity of the company's CO_2 emissions had been reduced by 3 per cent over its 2013 figures to 15.8 $kgCO_2e/modTEU$. This represents a significant decrease from the 20.2 $kgCO_2e/modTEU$ it reported for 2008. DP World's sustainability reporting has four main pillars: community, environment, marketplace, and people and safety. It has a dedicated sustainability advisory committee that sets development plans and policy and a sustainability champion team to identify best practices.

Other voluntary measures exist whereby a port may report upon its environmental impact. For instance, in Europe, the Port Environmental Review System, implemented by the European Sea Ports Organization, provides a series of commitments for a port authority to undertake, for example, the publication of a periodical report on the state of the environment, the monitoring of a series of environmental indicators, and the like. Another regional measure, which can be applied to ports, is the Hawkama Environment, Social and Governance Index for the Middle East and North Africa region. The Hawkama Index was developed in cooperation with Standard and Poor's reporting agency, with the support of the International Finance Corporation. The index provides an incentive to listed companies in these emerging markets to pursue sustainable business practices through improved environmental and socially responsible operations, as well as enhanced corporate governance systems.

Monitoring emissions and reporting on them with a view to reducing them over time requires the implementation of practical measures. The repositioning of empty container trucks within a port is a wasteful practice that can contribute to its overall emissions without transporting any goods. A proper computer-based monitoring, planning and coordination system to reduce unnecessary repositioning could have a significant impact on emissions without the need to spend money on purchasing new equipment or retrofitting exiting equipment with newer technology. Such a system will be most effective and likely to lead to the greatest emission savings if it operates on a concept of shared ownership of vehicles. Just as for private cars, individual ownership of transport modes tends to mean low occupancy and poor utilization rates. Any concept that includes sharing space on transport to and from a local dry port to a sea port could reduce the amount of emissions in and around the port.

Cold ironing, also known as "alternative maritime power" or "onshore power", is the process of providing electrical power to a ship while at berth. This means the ship's engines can be turned off, thereby reducing fuel emissions, vibrations and noise. Cold ironing displaces power generation from the vessel to a centralized power source, usually a power grid, which is generally more energy-efficient (GreenSync, 2015). Cold ironing does not eliminate emissions but transfers them to another location and may or may not be more polluting. The spread of ultra-low sulphur fuel and exhaust gas scrubbers have made significant air quality improvements around ports and coastal zones in recent years and has led some commenators to suggest that cold ironing may become obsolete (*The Maritime Executive*, 2015). However, cold ironing has the advantage that it can reduce noise and vibration emissions that cannot be eliminated by burning alternative fuels or by adopting exhaust capture solutions. In the European Union, directive 2014/94/EU obliges member States to implement alternative infrastructure networks such as shoreside power technology by December 2025. For shipowners, switching to cold ironing may prolong the life of a ship's equipment but will incur upfront funding in the form of electricity bills that may be higher than the cost of fuel oils. Marine diesel is usually purchased free of tax, but whether onshore electricity will carry the same advantage depends upon the national

Government. There is no international uniform voltage and frequency requirement when it comes to plugging in ships to national grids. Some ships use 220 volts at 50 Hz or 60 Hz, while others use 110 volts.

2. Social challenges

The main social challenges facing ports today include safety, security and reliability: safety, in terms of ensuring that employees or the general public are not injured; security, in respect of preventing dangerous or illegal goods from being smuggled into or out of ports; and reliability, in ensuring that the port is resilient enough to be able to continue at optimum performance levels regardless of any unwanted human or natural disturbance. Labour issues are, however, perhaps the most difficult of all issues to overcome. Dock work has traditionally been labour intensive, but increasingly labour-saving technologies are being introduced. The reform process usually starts with a port privatization process, of which retrenchment of labour is often a key feature. Any reduction in a workforce can cause considerable discontent both for the remaining workers and the larger community that is reliant on the dock workers' salaries. Yet in many places dock workers are employed under a protective status that limits access to the labour market to approved persons only. In Europe, there has long been an attempt to harmonize dock workers throughout the European Union, but as yet no clear-cut solution exists (Verhoeven, 2011). In 2014, dock workers in the Port of Piraeus protested about working conditions that included 16-hour working shifts (Vassilopoulos, 2014). In 2014 and 2015 in the United States, discussions between the International Longshore and Warehouse Union and the Pacific Maritime Association lasted months and led to severe traffic disruption to vessels entering and leaving the country's 29 west coast ports (Vekshin, 2015). In the port of Callao, Peru, a new system designed to automate the roster of shift workers met with protests resulting in the closure of the port's main container terminal (*Lloyd's List – Daily Briefing*, 2015). The challenge for Governments and port operators is in determining how to meet the demands of increased automation and yet still provide valued employment. Deregulation, which often precedes port privatization, can, however, lead to higher wages for those that remain in the industry. Research has found that the real (adjusted for inflation) hourly and weekly wages of United States union dockworkers increased by 14.3 per cent and 15.3 per cent, respectively, in the post-deregulation period (Talley, 2009).

3. Conclusions

With increased volumes, greater efficiencies and profits are materializing for terminal operators but not necessarily for port authorities. The immediate challenge for ports is not only adapting to these increased volumes but attending to global issues such as climate change mitigation and adaption. Increased automation is both helping and hindering this process. While human labour per se produces no harmful emissions, it is increasingly being replaced by automated machines that, while on the one hand increase terminal and port efficiency and may help lower transport costs, yet on the other hand tend to increase harmful emissions within the port area. The challenge for policymakers is to achieve the right policy mix that benefits both industry and society.

REFERENCES

ACS–AEC (2015). Trade facilitation: Port development and operations efficiency. Available at http://www.acs-aec.org/index.php?q=press-center/releases/2015/trade-facilitation-port-development-and-operations-efficiency (accessed 22 September 2015).

Bofinger HC, Cubas D and Briceno-Garmendia C (2015). OECS ports: An efficiency and performance assessment. Policy research working paper No. 7162. World Bank Group.

Cetin CK (2015). Port and logistics chains: Changes in organizational effectiveness. In: Song DW and Panayides P, eds., *Maritime Logistics: A Guide to Contemporary Shipping and Port Management.* Second edition. Kogan Page. London.

Channel News Asia (2015). Thailand denies Kra Canal deal. Available at http://www.channelnewsasia.com/news/asiapacific/thailand-denies-kra-canal/1856758.html (accessed 22 September 2015).

Data in Motion (2015). The JOC launches a new tool to benchmark port productivity. Available at https://pierstransportation.wordpress.com/2013/02/07/the-joc-launches-a-new-tool-to-benchmark-port-productivity/ (accessed 22 September 2015).

Doerr O (2011). Sustainable port policies. *Bulletin FAL*. 299(7). Available at http://repositorio.cepal.org/bitstream/handle/11362/36271/FAL-299-WEB-ENG_en.pdf?sequence=1 (accessed 17 September 2015).

Drewry (2014a). *Global Container Terminal Operators – Annual Review and Forecast 2014*. London.

Drewry (2014b). Container terminal capacity and performance benchmarks. October. Available at http://www.drewry.co.uk/publications/view_publication.php?id=425 (accessed 17 September 2015).

Drewry (2015). *Container Insight*. 3 May. Available at http://ciw.drewry.co.uk/release-week/2014-20/ (accessed 22 September 2015).

Economic and Social Commission for Asia and the Pacific (1992). *Assessment of the Environmental Impact of Port Development*. New York. Available at http://www.unescap.org/resources/assessment-environmental-impact-port-development-guidebook-eia-port-development (accessed 22 September 2015).

Fitzgerald WB, Howitt OJA and Smith IJ (2011). Greenhouse gas emissions from the international maritime transport of New Zealand's imports and exports. *Energy Policy*. 39(3):1521–1531.

Gracie C (2015). Wang Jing: The man behind the Nicaragua canal project. BBC News. Available at http://www.bbc.com/news/world-asia-china-31936549 (accessed 21 September 2015).

GreenSync (2015). Cold ironing within port's embedded networks. Available at http://www.greensync.com.au/cold-ironing-within-ports-embedded-networks/ (accessed 22 September 2015).

Hui-huang T (2015). A comparative study on pollutant emissions and hub-port selection in Panama Canal expansion. *Maritime Economics & Logistics*. 17(2).

JOC (2013). Introducing JOC port productivity. Available at http://www.joc.com/port-news/port-productivity/introducing-joc-port-productivity_20130201.html (accessed 14 September 2015).

IMO (2009). *Second IMO GHG 2009*. London. Available at http://www.imo.org/en/OurWork/Environment/PollutionPrevention/AirPollution/Documents/GHGStudyFINAL.pdf (accessed 22 September 2015).

Lam JSL and Notteboom T (2014). The greening of ports: A comparison of port management tools used by leading ports in Asia and Europe. *Transport Reviews*. 34(2).

Lloyd's List – Daily Briefing (2015). 5 June. Available at http://www.lloydslist.com/ll/daily-briefing/?issueDate=2015-06-05&expandId=462699 (accessed 22 September 2015).

Merk O (2014). Shipping emissions in ports. Discussion paper 2014-20. International Transport Forum. Paris.

Pasetti A (2015). The only way is up as APMT keeps faith with box terminal operator Global Ports. 1 August. *The Loadstar*. Available at http://theloadstar.co.uk/global-ports-apm-terminals-ap-moller-maersk/ (accessed 22 September 2015).

PSA (2014). *Annual Report 2014*. Available at https://www.globalpsa.com/ar/ (accessed 22 September 2015).

Suárez-Alemán A, Morales Sarriera J, Serebrisky T and Trujillo L (2015). When it comes to container port efficiency, are all developing regions equal? Inter-American Development Bank working paper 568. January. Available at http://idbdocs.iadb.org/wsdocs/getdocument.aspx?docnum=39360687 (accessed 22 September 2015).

Talley WK (2009). *Port Economics*. Routledge. London.

The Maritime Executive (2015). Is cold ironing redundant now? Available at http://www.maritime-executive.com/features/is-cold-ironing-redundant-now (accessed 22 September 2015).

van Dyck GK (2015). Assessment of port efficiency in West Africa using data envelopment analysis. *American Journal of Industrial and Business Management*. 5(4):208–218.

van Marle G (2015). Measuring port performance. *LongRead*. 1. June. Available at http://theloadstar.co.uk/wp-content/uploads/The-Loadstar-LongRead-Port-productivity1.pdf (accessed 22 September 2015).

Vassilopoulos J (2014). Dock workers at Piraeus Port, Greece end strike. World Socialist Web Site. Available at https://www.wsws.org/en/articles/2014/07/23/dock-j23.html (accessed 22 September 2015).

Vekshin JN (2015). United States West Coast port employees agree to deal. Available at http://www.bloomberg.com/news/articles/2015-02-20/west-coast-port-talks-said-to-intensify-as-perez-raises-pressure (accessed 22 September 2015).

Verhoeven P (2011). Dock labor schemes in the context of EU law and policy. *European Research Studies*. 14(2):149.

Vietnam Briefing (2015). Privatization of Viet Nam's port infrastructure to boost efficiency and lower prices. Available at http://wwjw.vietnam-briefing.com/news/privatization-vietnams-port-infrastructure-boost-efficiency-prices.html/ (accessed 22 September 2015).

Villalbaa G and Gemechub ED (2011). Estimating GHG emissions of marine ports – The case of Barcelona. *Energy Policy*. 39(3):1363–1368.

Westports (2015). Our milestones. Available at http://www.westportsmalaysia.com/About_Us-@-Our_Milestones.aspx (accessed 21 September 2015).

Yu A (2015). Chinese ports handled 202 million TEU in 2014. *Journal of Commerce*. 4 May. Available at http://www.ihsmaritime360.com/article/17726/chinese-ports-handled-202-million-teu-in-2014 (accessed 14 September 2015).

WTO (2014). Agreement on Trade Facilitation. Article 14: Categories of provisions. WT/L931. 15 July. Available at http://www.wto.org/english/news_e/news14_e/sum_gc_jul14_e.htm (accessed 9 September 2015).

WTO (2015). Doha development agenda. Available at http://www.wto.org/english/thewto_e/coher_e/mdg_e/dda_e.htm (accessed on 17 September 2015).

ENDNOTES

[1] The concept of "environmental space" describes the total amount of non-renewable resources, agricultural land and forests that can be used globally without impinging on access by future generations to the same resources. For one explanation of the environmental space concept, see the European Environment Agency: http://www.eea.europa.eu/publications/92-9167-078-2/page003.html (accessed 22 September 2015).

5
LEGAL ISSUES AND REGULATORY DEVELOPMENTS

In 2014, important regulatory developments in the field of transport and trade facilitation included the adoption of the International Code for Ships Operating in Polar Waters (Polar Code), expected to enter into force on 1 January 2017, as well as a range of regulatory developments relating to maritime and supply chain security and environmental issues.

To further strengthen the legal framework relating to ship-source air pollution and the reduction of greenhouse gas (GHG) emissions from international shipping, several regulatory measures were adopted at IMO, and the third IMO GHG Study 2014 was finalized. Also, guidelines for the development of the Inventory of Hazardous Materials required under the 2010 International Convention on Liability and Compensation for Damage in Connection with the Carriage of Hazardous and Noxious Substances by Sea (HNS Convention) – which, however, is not yet in force – were adopted, and further progress was made with respect to technical matters related to ballast water management, ship recycling, and measures helping to prevent and combat pollution of the sea from oil and other harmful substances.

Continued enhancements were made to regulatory measures in the field of maritime and supply chain security and their implementation, including the issuance of a new version of the World Customs Organization (WCO) Framework of Standards to Secure and Facilitate Global Trade (SAFE Framework) in June 2015, which includes a new pillar 3: "Customs-to-other government and inter-government agencies". As regards suppression of maritime piracy and armed robbery, positive developments were noted in the waters off the coast of Somalia and the wider western Indian Ocean. However, concern remains about the seafarers still being held hostage. A downward trend of attacks in the Gulf of Guinea was also observed, indicating that international, regional and national efforts are beginning to take effect.

A. IMPORTANT DEVELOPMENTS IN TRANSPORT LAW

1. Adoption of the International Code for Ships Operating in Polar Waters

The International Code for Ships Operating in Polar Waters (Polar Code), a new mandatory instrument establishing safety and environmental rules that are applicable to both Arctic and Antarctic shipping, was recently adopted at IMO. As noted in its preamble, the Polar Code "has been developed to supplement existing IMO instruments in order to increase the safety of ships' operation and mitigate the impact on the people and environment in the remote, vulnerable and potentially harsh polar waters". Part I of the Polar Code, which establishes safety-related requirements, along with associated amendments[1] to make it mandatory under the International Convention for the Safety of Life at Sea (SOLAS), was adopted in November 2014 by the IMO's Maritime Safety Committee (MSC) in response to the increasing numbers of ships operating in Arctic and Antarctic waters. Part I of the Polar Code addresses the safety of shipping in polar waters and identifies measures required over and above standard shipping regulations to ensure that ships can operate safely under the difficult conditions in these waters. Part II of the Code, which addresses the prevention of pollution from shipping, along with associated amendments to make it mandatory under MARPOL, was adopted by the IMO Marine Environment Protection Committee (MEPC) in May 2015.

The complete Polar Code is expected to enter into force on 1 January 2017 through the tacit acceptance procedure.[2] Thus, it will apply to new ships constructed on or after 1 January 2017. Ships constructed before that date will need to meet the relevant requirements of the Code by the first intermediate or renewal survey, whichever occurs first, after 1 January 2018.

Background

Oceans play a central role in helping regulate the climate, absorbing CO_2, providing food and nutrition and supporting livelihoods. However, ocean resources and services are exposed to threats including those associated with GHG emissions and air pollution; ocean acidification; illegal, unreported and unregulated fishing; and marine pollution. As highlighted by the United Nations Secretary-General in his remarks on the occasion of World Oceans Day 2015, oceans "are an essential element in our emerging vision for sustainable development, including the new set of sustainable development goals now being prepared to guide the global fight against poverty for the next 15 years" (United Nations Environment Programme, 2015). Noting that adopting agreements on climate change and ending poverty "will demand that [Governments] look at the essential role of [the] world's oceans", he called for a commitment to using "the gift of the oceans peacefully, equitably and sustainably for generations to come".[3]

Polar waters deserve particular attention due to special conditions that make them more vulnerable to the impacts of commercial shipping such as, for instance, ship-source pollution. Large populations of wildlife in polar areas are completely dependent on the living resources in the oceans, and even a small oil spill may have devastating consequences for biodiversity and ecosystem health. Also, oil and chemical discharges and spills persist for much longer in the colder polar waters, thus having a greater impact on wildlife and on the livelihoods of people in these areas, both directly and indirectly, through the impact on food.[4] At the same time, ships operating in polar waters and people aboard them are also exposed to a number of unique risks due, particularly, to the presence of large ice concentrations, poor weather conditions, extreme cold temperatures, remoteness and associated difficulties. Problems faced include, for instance, structural risks and difficulties in ships' operations, reduced efficiency of ships' machinery and equipment, lack of updated charts and navigation aids, difficulty in carrying out clean-up operations and difficulty or lack of availability of assistance from other ships in case of casualty.[5]

While polar shipping poses distinct operational challenges, the potential for shipping through Arctic waters has increased significantly in recent years. As a result of global warming and increasing rates of Arctic sea ice loss, new shipping lanes have opened up, mainly in summer, which might considerably shorten the shipping distances between Europe and Asia as compared to traditional routes, in particular those transiting through the Panama Canal. Thus, if the potential Arctic Sea lanes were fully open for traffic, savings on distance, time and costs – as well as fuel – could be achieved.[6] For instance, a navigable North-West Passage offers a route between Tokyo and New York that is 7,000 kilometres shorter than the route through the Panama Canal. Taking into account canal fees, fuel costs and other relevant factors that determine freight rates, the new trade lanes could cut the cost of a single voyage by a large container ship by as much as 20 per cent (Bergerson, 2008). Potential savings could be even greater for megaships unable to fit through the Panama and Suez Canals and currently sailing around the Cape of Good Hope and Cape Horn. It has been suggested that these potential shortcuts could foster greater competition with existing routes, including through a reduction in transport costs, thereby promoting trade and international economic integration (Wilson et al., 2004).

While the economic viability of trade along these new shipping lanes remains to be more fully explored, the volume and diversity of polar shipping is predicted to grow over the coming years. Challenges related to commercial shipping in an area which is both environmentally sensitive and operationally difficult need to be addressed, including through regulatory measures that serve to ensure that polar shipping develops in a safe and sustainable way, protecting both the safety of life at sea and the sustainability of the polar environments.[7] Communities living in the polar areas may require capacity-building assistance to respond to the challenges associated with increasing commercial shipping in the region.[8]

Regulatory framework for polar shipping

The framework instrument governing the rights and responsibilities of nations in their use of oceans and the regulation of shipping is the 1982 United Nations Convention on the Law of the Sea (UNCLOS), whose provisions also apply in polar areas, with respect to the jurisdictional status of polar waters and international straits, maritime boundaries, navigational rights and freedoms, as well as coastal and port State control.[9] Particularly relevant is article 234 of the Convention entitled "Ice-covered areas" providing that "Coastal States have the right to adopt and enforce non-discriminatory laws and regulations for the prevention, reduction and control of marine pollution from vessels in ice-covered areas within the limits of the exclusive economic zone". Such safety and environmental standards may be adopted by States either individually, through their national legislations, or collectively, through conventions and other instruments negotiated at international organizations, or regionally. The provisions of UNCLOS are supplemented by a broader regulatory framework, consisting of a number of international conventions and other legal instruments negotiated and adopted mainly at IMO and the International Labour Organization (ILO), which deal with a wide range of safety, environmental and seafarers' issues. Many of these legal instruments are widely accepted by States and their provisions are applicable generally, including in the polar areas, for States that are parties to them. Main conventions that establish mandatory rules and regulations include SOLAS, MARPOL and the Maritime Labour Convention, 2006 (MLC).

SOLAS[10] is the main convention in the area of shipping safety, establishing international safety standards for the construction, machinery, equipment and operation of ships.[11] As regards marine environmental protection, the main convention is MARPOL,[12] which aims at the prevention of pollution of the marine environment by ships from operational or accidental causes; six technical annexes specifically deal with prevention and control of pollution by oil (annex I); noxious liquid substances carried in bulk (annex II); harmful substances carried by sea in packaged form (annex III); sewage from ships (annex IV); garbage from ships (annex V); and air pollution from ships (annex VI).[13] Also worth noting in the context of pollution control and navigational safety is the Nairobi Wreck Removal Convention, 2007, which entered into force on 14 April 2015, key features of which were highlighted in last year's *Review of Maritime Transport* (UNCTAD, 2014a).[14] The regulation of seafarers' issues also plays an important role, in particular given that seafarers' working and living conditions can affect not only their own well-being and safety, but also the safety of ships and the protection of the marine environment from pollution. The MLC,[15] consolidating more than 68 international labour standards relating to seafarers, is the main international instrument that addresses seafarers' working and living conditions. Conditions in relation to seafarer competency, training and other matters related to ensuring the safety of ships and the people on board are mainly addressed through STCW and SOLAS. Amendments to the STCW and the STCW Code, adopted in Manila in June 2010, included "Training guidance for personnel on ships operating in ice-covered waters", and "Measures to ensure the competency of masters and officers of ships operating in polar waters".

The development of specific rules dedicated to polar shipping, which complement the general instruments on maritime safety and marine environmental protection mentioned above, began in the early 1990s, initially with a regulatory focus on the Antarctic area. For example, IMO designated the waters south of

60 degrees south latitude as an Antarctic Special Area[16] under MARPOL, for annex I (Prevention and control of pollution by oil),[17] annex II (Noxious liquid substances)[18] and annex V (Garbage from ships).[19] In addition, an amendment to MARPOL annex I prohibited the carriage and use of heavy fuel oils in Antarctic waters.[20] Moreover, under the Antarctic Treaty System,[21] much stricter environmental standards for vessel wastewater and garbage (including food waste) discharge were put in place for the Antarctic.[22] Beginning in the 2000s, some of the regulatory focus shifted to the Arctic and in 2002, IMO approved voluntary "Guidelines for ships operating in Arctic ice-covered waters" (IMO, 2002). These provide requirements additional to those already contained in SOLAS and MARPOL, taking into account the specific climatic conditions in Arctic waters, in order to meet appropriate standards of maritime safety and pollution prevention. With scientific findings increasingly suggesting a greater potential for commercial shipping through newly opened shipping lanes, in December 2009 voluntary guidelines for ships operating in polar waters were adopted, applicable to both Arctic and Antarctic areas (IMO, 2009). In February 2010, work commenced at IMO to turn these guidelines into a legally binding instrument (the Polar Code) that would help ensure environmental protection and foster the sustainable development of shipping in polar waters both in the Arctic and the Antarctic.

Key features of the Polar Code

As stated in its introduction, the goal of the Polar Code is to "provide for safe ship operation and the protection of the polar environment by addressing risks present in polar waters and not adequately mitigated by other instruments of the IMO". The Code acknowledges that polar water operation may impose additional demands on ships, their systems and their operation, beyond existing requirements of SOLAS, MARPOL and other relevant binding IMO instruments. It also acknowledges that "while Arctic and Antarctic waters have similarities, there are also significant differences. Hence, although the Code is intended to apply as a whole to both Arctic and Antarctic, the legal and geographical differences between the two areas have been taken into account".[23]

The Polar Code consists of two substantive parts dealing, respectively, with safety (part I) and pollution prevention (part II), together with an introduction which contains mandatory provisions applicable to both parts I and II. Mandatory provisions on safety measures are set out in part I-A, while related recommendations are set out in part I-B. Mandatory provisions on pollution prevention are contained in part II-A, again supplemented by related recommendations, set out in part II-B.

Part I-A of the Polar Code, entitled "Safety measures", includes chapters on: general issues; polar water operational manuals; ship structure; subdivision and stability; watertight and weathertight integrity; machinery installations; fire safety and protection; life-saving appliances and arrangements; safety of navigation; communication; voyage planning; staffing and training. Each of these chapters sets out goals, functional requirements and relevant regulations. Part I-B establishes "Additional guidance regarding the provisions of the introduction and part I-A".

Part II-A of the Polar Code, entitled "Pollution prevention measures" includes chapters on: prevention of oil pollution; control of pollution from noxious liquid substances in bulk; prevention of pollution by harmful substances carried by sea in packaged form; prevention of pollution by sewage from ships; and prevention of pollution by garbage from ships. Part II-B contains "Additional guidance to part II", including also guidance on other environmental conventions and guidelines, more specifically related to ballast water management and anti-fouling coatings.

The Polar Code will apply to passenger ships and cargo ships of 500 GT and above, and covers the full range of shipping-related matters relevant to navigation in waters surrounding the two poles. It will require ships intending to operate in Arctic and Antarctic waters to undergo an assessment, taking into account the anticipated range of operating conditions and hazards the ship may encounter in the polar waters, and apply for a Polar Ship Certificate, which would classify the vessel according to the categories below:

- Category A ship: Designed for operation in at least medium first-year ice which may contain old ice inclusions (polar class 1 to 5 or equivalent);
- Category B ship: Designed for operation in at least thin first-year ice which may contain old ice inclusions (polar class 6 and 7 or equivalent);
- Category C ship: Designed for operation in open water or in ice conditions less severe than those in categories A and B.

Ships will also need to carry a Polar Water Operational Manual to provide the owner, operator, master and crew with sufficient information regarding the ship's operational capabilities and limitations to support their decision-making process.

CHAPTER 5: LEGAL ISSUES AND REGULATORY DEVELOPMENTS

Key elements of part II of the Code regarding environmental issues include:

- Discharge into the sea of oil or oily mixtures from any ship is prohibited. Oil fuel tanks must be separated from outer shells;
- Discharge into the sea of noxious liquid substances, or mixtures containing such substances, is prohibited;
- Discharge of sewage is prohibited unless performed in line with MARPOL annex IV and requirements in the Polar Code;
- Discharge of garbage is restricted and only permitted in accordance with MARPOL annex V and requirements in the Polar Code.

In addition, some non-mandatory guidance is provided regarding measures to address, inter alia, potential threats from invasive species introduced via ballast water discharges[24] or through hull fouling (part II-B).

Part II does not appear to provide significant additional protection for Antarctic waters because there are already a number of regulations in place that prohibit the discharge of oil, noxious liquids and various forms of garbage in those waters. It will, however, improve the protection of Arctic waters from the discharge of these wastes, bringing the requirements for Arctic waters more in line with the existing protections in place for Antarctic waters.

B. REGULATORY DEVELOPMENTS RELATING TO THE REDUCTION OF GREENHOUSE GAS EMISSIONS FROM INTERNATIONAL SHIPPING AND OTHER ENVIRONMENTAL ISSUES

1. Reduction of greenhouse gas emissions from international shipping and energy efficiency

During the sixty-seventh and sixty-eighth sessions of MEPC,[25] States continued to focus on the reduction of CO_2 emissions from international shipping, including through improving ships' design and size, better speed management, and other operational measures, to reduce ships' consumption of fuel. The issue of possible market-based measures for the reduction of GHG emissions from international shipping was not addressed, as further discussions on this had been postponed to a future session.[26] It should be recalled that a new set of technical and operational measures to increase energy efficiency and reduce emissions of GHGs from international shipping had been adopted in 2012 (IMO, 2011, annex 19).[27] This package of measures, introducing the EEDI for new ships and the SEEMP for all ships, was added by way of amendments to MARPOL annex VI through the introduction of a new chapter 4 entitled "Regulations on energy efficiency for ships", which entered into force on 1 January 2013. Guidelines and unified interpretations to assist in the implementation of this set of technical and operational measures were subsequently adopted at IMO in 2012, 2013 and 2014. In addition, a "Resolution on promotion of technical cooperation and transfer of technology relating to the improvement of energy efficiency of ships" was adopted in May 2013, and a new study to provide an update to the IMO 2009 GHG emissions estimate for international shipping was completed in 2014. Information about relevant deliberations and outcomes during the sixty-seventh and sixty-eighth sessions of MEPC is presented below.

Reduction of greenhouse gas emissions from international shipping

An important development during the sixty-seventh session of MEPC was the approval of the third IMO GHG study 2014 (IMO, 2014a). The study provides updates to earlier estimates for GHG emissions from ships contained in the second IMO GHG study (2009). The third IMO GHG study estimates that international shipping emitted 796 million tons of CO_2 in 2012, compared to 885 million tons in 2007. This represented 2.2 per cent of the global emissions of CO_2 in 2012, compared to 2.8 per cent in 2007.[28]

The main findings of the study as regards scenarios for 2012–2050 include the following:

- Maritime CO2 emissions are projected to increase significantly. Depending on future economic and energy-related developments, this study's "business as usual" scenarios project an increase by 50 to 250 per cent in the period to 2050. Further action on efficiency and emissions can mitigate the emissions growth, although all scenarios but one project emissions in 2050 to be higher than in 2012;

- Among the different cargo categories, demand for transport of unitized cargos is projected to increase most rapidly in all scenarios;
- Emission projections demonstrate that improvements in fuel efficiency are important to mitigate emission increases. However, even modelled improvements with the greatest energy savings do not yield a downward trend. Compared to regulatory or market-driven improvements in efficiency, changes in the fuel mix have a limited impact on GHG emissions, assuming that fossil fuels remain dominant;
- Most other emissions increase in parallel with CO_2 and fuel, with some notable exceptions. Methane emissions are projected to increase rapidly (albeit from a low base) as the share of LNG in the fuel mix increases. Emissions of NO_x may increase at a lower rate than CO_2 emissions as a result of tier II and tier III engines entering the fleet. Emissions of PM show an absolute decrease until 2020, and SO_x continue to decline through 2050, mainly because of MARPOL annex VI requirements on the sulphur content of fuels.

At its sixty-eighth session, MEPC considered a submission from one member State calling for a quantifiable reduction target for GHG emissions from international shipping, consistent with keeping global warming below 1.5°C, and for agreement on the measures necessary to reach that target (IMO, 2015a, annex 25).[29] During the discussion, speakers acknowledged the importance of the issue raised and of the establishment of emissions reporting for international shipping as a matter of priority. They also recognized that, despite the measures already taken by IMO regarding the reduction of emissions from ships, more could be done. However, MEPC took the view that the priority at this stage should be to continue its current work, in particular to focus on further reduction of emissions from ships through the finalization of a data collection system for fuel consumption.

Energy efficiency for ships

MEPC continued its work on further developing guidelines to assist member States in the implementation of the mandatory energy-efficiency regulations for international shipping. At its sixty-seventh and sixty-eighth sessions, MEPC in particular adopted:

- "2014 Guidelines on survey and certification of the Energy Efficiency Design Index" (IMO, 2014b, annex 5);[30]
- "Amendments to the 2013 Interim Guidelines for determining minimum propulsion power to maintain the manoeuvrability of ships in adverse conditions" (IMO, 2014b, annex 6);[31]
- "Amendments to the 2014 Guidelines on survey and certification of the Energy Efficiency Design Index" (IMO, 2015a, annex 6), and endorsed their application from 1 September 2015, at the same time encouraging earlier application;
- "Amendments to the 2013 Interim Guidelines for determining minimum propulsion power to maintain the manoeuvrability of ships in adverse conditions" (IMO, 2015a, annex 7);[32]
- "Amendments to the 2014 Guidelines on the method of calculation of the attained EEDI for new ships" (IMO, 2015a, annex 8).

MEPC also considered a progress report from the intersessional correspondence group established at its previous session to review the status of technological developments relevant to implementing phase 2 of the EEDI regulations[33] and re-established the correspondence group to further the work and submit an interim report to the sixty-ninth session of MEPC.

Further technical and operational measures for enhancing the energy efficiency of international shipping

In respect of a proposed data collection system for the fuel consumption of ships, which could be used, inter alia, to estimate CO_2 emissions, MEPC at its sixty-eighth session agreed that text prepared by the intersessional correspondence group[34] should be further developed in the form of full language for the data collection system, which could be readily used for voluntary or mandatory application of the system. The core elements of the data collection system include data collection by ships, functions of flag States in relation to data collection and the establishment of a centralized database by IMO. According to the proposed text, data would be collected for ships of 5,000 GT and above and include the ship identification number, technical characteristics, total annual fuel consumption by fuel type and in metric tons, and transport work and/or proxy data yet to be defined. The methodology for collecting the data would be outlined in the ship-specific SEEMP. Data, aggregated into an annual figure, would be reported by the shipowner/operator to the administration (flag State), which would submit the data to IMO for inclusion in a database, with access restricted to member States only, and with data anonymized to the extent that the identification of a specific ship would not be possible.

MEPC noted that one purpose of the data collection system was to analyze energy efficiency, and for this analysis to be effective some transport work data needed to be included. However, at this stage the appropriate parameters have not been identified. MEPC recommended that an intersessional working group be held to further consider transport work and/or proxies for inclusion in the data collection system, further consider the issue of confidentiality, and consider the development of guidelines identified in the text.

Matters concerning the United Nations Framework Convention on Climate Change

MEPC noted a document on the outcomes of the United Nations Climate Change Conferences held in Lima in December 2014 and in Geneva in February 2015 (IMO, 2015b). It requested the IMO secretariat to continue its cooperation with the UNFCCC secretariat, and to bring the outcome of the work of IMO to the attention of appropriate UNFCCC bodies and meetings, as necessary.

2. Ship-source pollution and protection of the environment

(a) Air pollution from ships

MEPC continued its work on developing regulations to reduce emissions of other toxic substances from burning fuel oil, particularly NO_x and SO_x. Together with CO_2, these significantly contribute to air pollution from ships, and are covered by annex VI of MARPOL,[35] amended in 2008 to introduce more stringent emission controls.

During its sixty-eighth session, MEPC considered a number of amendments to existing guidance and other issues related to air pollution measures, and:

- Adopted "2015 Guidelines for exhaust gas cleaning systems" (IMO, 2015a, annex 1). These relate to certain aspects of emission testing regarding measurements of CO_2 and SO_2, clarification of the wash water discharge pH limit testing criteria, and the inclusion of a calculation-based methodology for verification as an alternative to the use of actual measurements;

- Approved the Bond et al.[36] definition of black carbon for international shipping as a distinct type of carbonaceous material formed only in flames during combustion of carbon-based fuels. It is distinguishable from other forms of carbon and carbon compounds contained in atmospheric aerosol because of its unique combination and physical properties.

MEPC also noted that it was not possible at this stage to consider possible control measures to reduce the impact on the Arctic of emissions of black carbon from international shipping.[37]

Emissions of nitrogen oxides

As highlighted in previous issues of the *Review of Maritime Transport*, measures have been adopted at IMO that require ships to gradually produce NO_x emissions below the tier III level. Tier III limits are almost 70 per cent lower than those of the preceding tier II, thus requiring additional technology. During its sixty-seventh and sixty-eighth sessions, MEPC continued its consideration of issues related to progressive reductions in NO_x emissions from ship engines, and in particular:

- Adopted amendments to MARPOL annex VI (IMO, 2014b, annex 9), concerning regulation 2 (definitions), regulation 13 (NO_x) and the Supplement to the International Air Pollution Prevention Certificate, in order to include reference to gas as fuel and to gas-fuelled engines. These are expected to enter into force on 1 March 2016;

- Approved draft amendments to the NO_x Technical Code 2008 (testing of gas-fuelled engines and dual-fuel engines for the NO_x tier III strategy) (IMO, 2014b, annex 3);

- Approved draft amendments to MARPOL annex VI (record requirements for operational compliance with NO_x tier III ECAs) (IMO, 2014b, annex 4);

- Approved guidance on the application of regulation 13 of MARPOL annex VI tier III requirements to dual-fuel and gas-fuelled engines (IMO, 2015c);

- Adopted amendments to the 2011 guidelines addressing additional aspects of the NO_x technical code 2008 with regard to particular requirements related to marine diesel engines fitted with selective catalytic reduction systems (IMO, 2015a, annex 2);

- Agreed, for consistency and safety reasons, to proceed with the development of guidelines for the sampling and verification of fuel oil used on board ships.

Thus, tier III standards will apply to a marine diesel engine that is installed on a ship constructed on or after 1 January 2016 and which operates in the North American ECA or the United States Caribbean Sea ECA that are designated for the control of NO_x emissions. In addition, tier III standards will apply to installed marine diesel engines when operated in other ECAs that might be designated in the future for tier III NO_x control. They will apply to ships constructed on or after the date of adoption by MEPC of such an ECA, or a later date as may be specified in the amendment designating the NO_x tier III ECA.[38] Furthermore, tier III requirements will not apply to a marine diesel engine installed on a ship constructed prior to 1 January 2021 of less than 500 GT, of 24 metres or over in length, which has been specifically designed and is used solely for recreational purposes. These amendments are expected to enter into force on 1 September 2015. Requirements for the control of NO_x apply to installed marine diesel engines of over 130 kilowatt output power, and different levels (tiers) of control apply based on a ship's construction date. Outside ECAs designated for NO_x control, tier II controls, required for marine diesel engines installed on ships constructed on or after 1 January 2011, apply. While IMO tier III standards will come into force for ships constructed from 1 January 2016 onwards, it has been noted that retrofitting existing vessels with tier III technology, where possible, could significantly enhance fuel efficiency for existing fleets, thus reducing both emissions and operational costs (*The Ship Supplier*, 2014).

Sulphur oxide emissions

As reported in previous editions of the *Review of Maritime Transport*, with effect from 1 January 2012, MARPOL annex VI established reduced SO_x thresholds for marine bunker fuels, with the global sulphur cap reduced from 4.5 per cent (45,000 parts per million (ppm)) to 3.5 per cent (35,000 ppm), outside an ECA. The global sulphur cap is expected to be reduced further to 0.5 per cent (5,000 ppm) from 2020. Depending on the outcome of a review, to be completed by 2018, as to the availability of compliant fuel oil, this requirement could be deferred to 1 January 2025. Within ECAs where more stringent controls on sulphur emissions apply, the sulphur content of fuel oil must be no more than 1 per cent, falling to 0.1 per cent (1,000 ppm) from 1 January 2015.[39]

To meet these new guidelines, shipowners and operators are adopting a variety of strategies. These include switching to low-sulphur fuels, installing scrubbers and switching to LNG as fuel. However, implementing these strategies may be costly. For instance, the supply of low-sulphur marine gas oil remains a concern, and other distillate alternatives available are expensive. Installing scrubbers or exhaust gas SO_x cleaning systems on ships may cost $3 million–$5 million per scrubber, and LNG retrofitting is very expensive and not always feasible. Operators therefore risk being fined for breaching emission restrictions and some of them may, in the short term, choose to accept this situation (*IHS Maritime Technology*, 2014).

The 2010 guidelines for monitoring the worldwide average sulphur content of fuel oils supplied for use on board ships (IMO, 2010, annex I) provide for the calculation of a rolling average of the sulphur content for a three-year period. The rolling average based on the average sulphur contents calculated for the years 2012–2014 is 2.47 per cent for residual fuel and 0.13 per cent for distillate fuel (IMO, 2013, 2014c and 2015d).

At its sixty-eighth session, MEPC agreed that the IMO secretariat should initiate in 2015 a review of the availability of compliant fuel oil to meet the global requirement that the sulphur content of fuel oil used on board ships shall not exceed 0.50 per cent as from 1 January 2020.[40] The fuel oil availability review will be overseen by a steering committee[41] and a final report will be submitted to the seventieth session of MEPC in autumn 2016.

In addition, MEPC considered the report of a correspondence group established to consider possible quality control measures prior to fuel oil being delivered to a ship, and re-established it to further develop draft guidance on best practices for assuring the quality of fuel oil delivered for use on board ships; further examine the adequacy of the current legal framework in MARPOL annex VI for assuring the quality of fuel oil for use on board ships; and submit a report to the sixty-ninth session of MEPC.[42]

Other issues

During its sixty-seventh and sixty-eighth sessions, MEPC adopted the following amendments that are expected to enter into force on 1 March 2016:

- Amendments to MARPOL annex I (IMO, 2014b, annex 7) concerning regulation 43 on special requirements for the use or carriage of oils in the Antarctic area, and prohibiting ships from carrying heavy grade oil on board as ballast;

- Amendments to MARPOL annex III (IMO, 2014b, annex 8) concerning the appendix on criteria for the identification of harmful substances in packaged form.

MEPC also:

- Approved two sets of guidelines to assist in oil spill response, developed by the Subcommittee on Pollution Prevention and Response:
  - "Guidelines on international offers of assistance in response to a marine oil pollution incident" (IMO, 2015e, annex 13);[43]
  - "Guidelines for the use of dispersants for combating oil pollution at sea – Part III (Operational and technical sheets for surface application of dispersants)" (IMO, 2015e, annex 14).[44]
- Adopted "Amendments to regulation 12 of MARPOL annex I, concerning tanks for oil residues (sludge)" (IMO, 2014d). These expand on the requirements for discharge connections and piping to ensure oil residues are properly disposed of.

(b) Ballast water management

One of the major threats to biodiversity is the introduction of non-native species following the discharge of untreated ships' ballast water. Indeed, the introduction of harmful aquatic organisms and pathogens to new environments has been identified as one of the four greatest threats to the world's oceans.[45] Even though ballast water is essential to ensure safe operating conditions and stability for vessels at sea, it often carries with it a multitude of marine species, which may survive to establish a reproductive population in the host environment – becoming invasive, out-competing native species and multiplying into pest proportions. The proliferation of bioinvasions continues to increase in conjunction with the growth of seaborne trade, as approximately 10 billion tons of ballast water per year are transferred globally, with potentially devastating consequences. In February 2004, the International Convention for the Control and Management of Ships' Ballast Water and Sediments (BWM Convention) was adopted under the auspices of IMO to prevent, minimize and ultimately eliminate the risks to the environment, human health, property and resources arising from the transfer of harmful aquatic organisms carried by ships' ballast water from one region to another (for a review, see UNCTAD, 2011b).

During its sixty-seventh and sixty-eighth sessions, MEPC agreed to grant basic approval to six,[46] and final approval to four,[47] ballast water management systems that make use of active substances. In addition, at both sessions MEPC reviewed the status of the BWM Convention, which is close to fulfilling the remaining criteria (tonnage) for its entry into force. The Convention is set to enter into force twelve months after the date on which no fewer than 30 States, the combined merchant fleets of which constitute not less than 35 per cent of the GT of world merchant shipping, have become Parties to it. As of 30 June 2015, 44 States, representing 32.86 per cent of the world's merchant fleet GT, had become parties.[48]

MEPC also:

- Adopted "Resolution MEPC.252(67) on guidelines for port State control under the BWM Convention" (IMO, 2014b, annex 1);
- Adopted a "Plan of action for reviewing the guidelines for approval of ballast water management systems (G8)" (IMO, 2014b, annex 2);
- Adopted "Resolution MEPC.253(67) on measures to be taken to facilitate entry into force of the BWM Convention" (IMO, 2014b, annex 3);[49]
- Agreed on a "Road map for the implementation of the BWM Convention" (IMO, 2014e, annex 2). This explains that ships that install ballast water management systems approved in accordance with the current guidelines (G8), ("early movers"), should not be penalized;
- Developed "Draft amendments to regulation B-3 of the BWM Convention to reflect Assembly resolution A.1088(28) on application of the Convention", with a view to approval at the sixty-ninth session and consideration for adoption once the treaty enters into force. These will provide an appropriate timeline for ships to comply with the ballast water performance standard prescribed in regulation D-2 of the Convention;
- Received a progress report on a study on the implementation of the ballast water performance standard described in regulation D-2 of the Convention (IMO, 2015f).[50]

(c) Ship recycling

MEPC adopted the "2015 Guidelines for the development of the Inventory of Hazardous Materials" (IMO, 2015a, annex 17). The Inventory is required under the Hong Kong International Convention for the

Safe and Environmentally Sound Recycling of Ships, 2009 (Hong Kong Convention). The Convention is not yet in force and at 30 June 2015 only three States had ratified it. The Convention requires ratification by not less than 15 States to enter into force.

(d) Developments regarding the International Convention on Liability and Compensation for Damage in Connection with the Carriage of Hazardous and Noxious Substances by Sea, 1996, as amended by its 2010 Protocol

The issue of the entry into force of the 2010 HNS Convention was discussed by the Legal Committee of IMO at its 102nd session in April 2015. In particular, the mandate of the HNS Correspondence Group was extended to develop a publication entitled *Understanding the HNS Convention*,[51] another document entitled *HNS Scenarios* and a Legal Committee resolution that would help encourage States to implement the HNS Convention and take the necessary steps to bring it into force within a reasonable time.[52] As reported in previous editions of the *Review of Maritime Transport*, the HNS Convention, originally adopted in 1996, was amended in 2010 in an effort to overcome a number of perceived obstacles to ratification. However, despite the recognized importance of an international liability and compensation regime for HNS carried by sea (UNCTAD, 2012a), to date no State has ratified the HNS Convention as amended in 2010. As a result, it is not clear if and when the 2010 HNS Convention will enter into force and an important gap in the global liability and compensation framework remains.[53] It may be recalled that a comprehensive and robust international liability and compensation regime is in place in respect of oil pollution from tankers (the International Oil Pollution Compensation Fund regime),[54] while liability and compensation for bunker oil pollution from ships other than tankers is also effectively regulated in the International Convention on Civil Liability for Bunker Oil Pollution Damage, 2001 (Bunker Oil Pollution Convention).

(e) Liability and compensation for transboundary pollution damage resulting from offshore oil exploration and exploitation

It should be noted that the need for international regulation to provide for liability and compensation for transboundary pollution damage resulting from offshore exploration and exploitation activities was again considered by the IMO Legal Committee at its 102nd session. However, following discussion, the Legal Committe decided that there was currently no compelling need to develop an international convention and, as already agreed at its previous sessions, guidance on bilateral or regional agreements should continue to be developed (IMO, 2015g).

Offshore oil exploration poses particular technical, safety and operational challenges, which are increased in areas prone to earthquakes. Associated oil pollution incidents may have potentially devastating consequences, both in terms of economic loss and in terms of effects on marine biodiversity and ecosystem health, in particular in sensitive marine environments like the Arctic. While offshore oil exploration and exploitation is expected to grow in the future,[55] at present there is no international legal instrument to provide for liability and compensation in cases of accidental or operational oil spills.

With respect to liability and compensation for oil pollution from offshore platforms, recent developments related to the Deepwater Horizon disaster, one of the largest accidental marine oil spills in the world and the largest environmental disaster in United States history, are also worth noting. The disaster, which occurred in the Gulf of Mexico about 40 miles south-east of the Louisiana coast on 20 April 2010, was a result of the explosion, sinking of and subsequent massive oil spill from the Deepwater Horizon drilling rig, owned and operated by the company Transocean and drilling for British Petroleum (BP). The explosion killed 11 workers, injured 16 others and the total discharge was estimated at 4.9 million barrels (210 million United States gallons; 780,000 cubic metres).[56] In June 2015, more than five years after the disaster, BP's $18.7 billion settlement with various United States Government agencies of claims resulting from the explosion was announced. This was reportedly in addition to $29.1 billion in costs associated with the initial and ongoing clean-up operations and the settlement of civil claims brought by businesses damaged by the oil spill, bringing the final bill to approximately $50 billion.[57]

Key developments in summary

As the above overview indicates, during the year under review there were a number of regulatory initiatives and developments aimed at implementing sustainable development objectives and policies. These include, notably, the adoption of the Polar Code, which establishes mandatory provisions to

CHAPTER 5: LEGAL ISSUES AND REGULATORY DEVELOPMENTS

ensure ship safety and prevent environmental pollution in both Arctic and Antarctic waters. The Polar Code is expected to enter into force on 1 January 2017. In addition, the third IMO GHG study was finalized, providing an updated estimate of CO_2 emissions from international shipping over the period 2012–2050, and several regulatory measures were adopted at IMO to strengthen the legal framework relating to ship-source air pollution and the reduction of GHG emissions from international shipping. Guidelines for the development of the Inventory of Hazardous Materials, required under the 2010 HNS Convention, were adopted, and further progress was made with respect to technical matters related to the implementation of the 2004 BWM Convention and the 2009 Ship Recycling Convention. Following the decision of the IMO Legal Committee that there was no compelling need to develop an international convention, the important issue of liability and compensation for transboundary pollution resulting from offshore oil exploration and exploitation remains, for the time being, outside the ambit of international regulation.

C. OTHER LEGAL AND REGULATORY DEVELOPMENTS AFFECTING TRANSPORTATION

This section highlights some key issues in the field of maritime security and safety that may be of particular interest to parties engaged in international trade and transport. These include developments relating to maritime and supply chain security and maritime piracy.

1. Maritime and supply chain security

(a) World Customs Organization Framework of Standards to Secure and Facilitate Global Trade

As noted in previous editions of the *Review of Maritime Transport*, in 2005, WCO adopted the SAFE Framework with the objective of developing a global supply chain framework, while also recognizing the significance of a closer partnership between customs administrations and business. The SAFE Framework provides a set of standards and principles that must be adopted as a minimum threshold by national customs administrations, originally contained within two pillars: pillar 1, "Customs-to-customs network arrangements", and pillar 2, "Customs–business partnerships".[58] The SAFE Framework is a widely accepted instrument that serves as an important reference point for customs and for economic operators alike and has evolved over the years as a dynamic instrument.[59] It was first updated in 2007 to incorporate detailed provisions on the conditions and requirements for customs and AEOs (a status that reliable traders may be granted and that entails benefits in terms of trade-facilitation measures). In 2010, a SAFE Package was issued, which brought together all WCO instruments and guidelines that support implementation of the SAFE Framework, and in June 2012 a revised version of the SAFE Framework included new parts 5 and 6 in respect of coordinated border management and trade continuity and resumption. A new annex I for definitions, including definition of "high risk cargo", was also added.[60]

A revised version of the SAFE Framework was issued in June 2015 that includes a new pillar 3, "Customs-to-other government and inter-government agencies", aiming to foster closer cooperation between customs administrations and other government agencies involved in the international trade supply chain (WCO, 2015a). Pillar 3 foresees cooperation at three levels: cooperation within a Government; cooperation between and among Governments; and multinational cooperation. Standards for each of these areas have been developed to promote such cooperation through a multi-tiered approach. A number of tools have been developed by WCO that support this pillar, notably the Compendiums on Coordinated Border Management and Single Window, which are continually updated. Another important aspect of this SAFE version is the incorporation of standards for "pre-loading advance cargo information" in respect of air cargo to carry out a first layer of security risk analysis together with civil aviation authorities. It also includes definitions of "container" and "risk management". Furthermore, the instruments and tools related to risk management mentioned in technical specifications of standards 4 and 7 of pillar 1 and other relevant sections have been updated in view of the development of the WCO Risk Management Compendium, volumes 1 and 2.

An important feature of the SAFE Framework, AEOs[61] are private parties that have been accredited by national customs administrations as compliant with WCO or equivalent supply chain security standards. AEOs have to meet special requirements in respect of physical security of premises, hidden camera surveillance and selective staffing and recruitment policies. In return, AEOs are to be rewarded by way of trade-facilitation benefits, such as faster clearance of goods and fewer physical inspections. In recent years, a number of mutual recognition agreements (MRAs) of respective AEOs have been adopted by customs administrations, usually on a bilateral basis.[62] However, it is hoped that these bilateral agreements will, in due course, form the basis for multilateral agreements at the subregional and regional levels. As of June 2015, 37 AEO programmes had been established in 64 countries[63] and a further 16 countries plan to establish them in the near future.[64] Capacity-building assistance remains a vital part of the SAFE implementation strategy. During 2014 and the first quarter of 2015, AEO workshops under the WCO Columbus Programme, or under specific financial support, were organized in a number of countries.[65]

(b) Developments at the European Union level and in the United States

This subsection provides an update of developments in relation to existing maritime and supply chain security standards at the European Union level and in the United States, both important trade partners for many developing countries.

As regards the European Union, previous editions of the *Review of Maritime Transport* have provided information on the Security Amendment to the Community Customs Code,[66] which aims to ensure an equivalent level of protection through customs controls for all goods brought into or out of the European Union's customs territory.[67] Part of the changes to the Customs Code involved the development of common rules for customs risk management, including setting out common criteria for pre-arrival/pre-departure security risk analysis based on electronically submitted cargo information. From 1 January 2011, this advance electronic declaration of relevant security data has become an obligation for traders.[68]

Among the changes to the Customs Code was the introduction of provisions regarding AEOs. In this context, subsequent related developments – such as the recommendation for self-assessment of economic operators to be submitted together with their applications for AEO certificates,[69] and the issuance of a revised self-assessment questionnaire[70] to guarantee a uniform approach throughout all European Union member States – are also worth noting. The European Union has concluded six AEO MRAs with third countries, including major trading partners, and negotiations on another MRA are ongoing.[71]

The European Commission, on 21 August 2014, adopted a Communication on European Union Strategy and Action Plan for customs risk management: "Tackling risks, strengthening supply-chain security and facilitating trade" (European Commission, 2014a). The Strategy and Action Plan annexed to the Communication proposes a set of step-by-step actions to reach more coherent, effective and cost-efficient European Union customs risk management at the European Union's external borders (European Commission, 2014b).[72]

As regards developments in the United States, according to the United States Customs and Border Protection (CBP), more than 11 million maritime containers arrive at United States seaports each year. At land borders, another 11 million arrive by truck and 2.7 million by rail.[73] Programmes such as the CSI and the Customs–Trade Partnership against Terrorism (C–TPAT), in which representatives of the trade community participate, help to increase the security of trade along supply chains.[74] Within months of the 11 September 2001 attacks, CSI was established to address the threat to border security and global trade posed by the potential for terrorist use of a maritime container to deliver a weapon. CSI aims to ensure all containers that pose a potential risk are identified and inspected at foreign ports before they are placed on ships destined for the United States. Teams of CBP officers are stationed in foreign locations to work together with their host foreign government counterparts, in order to target and pre-screen containers through "non-intrusive inspection" and radiation detection technology, as early in the supply chain and as rapidly as possible without slowing down trade. Since the inception of CSI, a significant number of customs administrations have joined the programme, and CSI is now operational at 58 ports in North America, Europe, Asia, Africa, the Middle East, and Latin and Central America, pre-screening over 80 per cent of all maritime containerized cargo imported into the United States.[75]

Starting as a partnership in November 2001 with seven major importers from the United States and neighbouring countries as members, the C–TPAT currently includes more than 10,000 certified partners from the trade community. When joining the C–TPAT, companies sign an agreement to work with CBP to protect the supply chain, identify security gaps and implement specific security measures and best practices. Additionally, partners provide CBP with a security profile outlining the specific security measures the companies have in place. C-TPAT members are considered low risk and are therefore less likely to be examined. C-TPAT signed its first MRA in June 2007 and since then has signed similar arrangements with nine countries/territories and the European Union.[76]

As highlighted in the *Review of Maritime Transport 2009*, in January 2009 new requirements, known as the "10+2" rule, came into effect.[77] The rule requires both importers and carriers to submit additional information pertaining to cargo to CBP before the cargo is brought into the United States by vessel. Failure to comply with the rule could ultimately result in monetary penalties, increased inspections and delay of cargo.[78]

Also worth mentioning are the voluntary Importer Self Assessment programme, in place since June 2002, which provides the opportunity for interested importers who are participating members of C-TPAT to assume responsibility for monitoring their own compliance in exchange for benefits;[79] the recent Trusted Trader programme, already in the test phase, which aims to join and unify the existing C-TPAT and Importer Self Assessment programmes to integrate and streamline the processes of supply chain security and trade compliance within one partnership programme;[80] and the Proliferation Security Initiative, which aims to stop trafficking of weapons of mass destruction, their delivery systems, and related materials to and from those State and non-State actors which may be of concern regarding arms proliferation. In February 2004, the Proliferation Security Initiative was expanded to include law enforcement cooperation, and to date more than 100 countries around the world have endorsed it.[81]

In addition, the United States has coordinated and supported other international initiatives, including the expansion of the WCO SAFE Framework, by providing targeted countries with training and advisory support through programmes on capacity-building and export control and border security.[82]

(c) International Organization for Standardization

During the last decade, the International Organization for Standardization (ISO) has been actively engaged in matters of maritime transport and supply chain security. Shortly after the release of the International Ship and Port Facility Security Code (ISPS Code),[83] and to facilitate its implementation by the industry, the ISO technical committee ISO/TC 8 published ISO 20858:2007, "Ships and marine technology – Maritime port facility security assessments and security plan development". Also relevant is the development of the ISO 28000 series of standards "Security management systems for the supply chain", which are designed to help the industry successfully plan for, and recover from, any ongoing disruptive event (box 5.1 details the current status of the ISO 28000 series). The core standard in this series is ISO 28000:2007, "Specification for security management systems for the supply chain", which serves as an umbrella management system that enhances all aspects of security – risk assessment, emergency preparedness, business continuity, sustainability, recovery, resilience and/or disaster management – whether relating to terrorism, piracy, cargo theft, fraud or many of the other security disruptions. The standard also serves as a basis for AEO and C–TPAT certifications. Various organizations adopting such standards may tailor an approach compatible with their existing operating systems. The standard ISO 28003:2007, published and in force since 2007, provides requirements for providing audits and certification to ISO 28000:2007.

The recent ISO 28007-1:2015, published in April 2015, cancels and replaces ISO/PAS 28007:2012 that provided guidelines containing additional sector-specific recommendations, which companies or organizations that comply with ISO 28000 can implement before they provide privately contracted armed security personnel on board ships. However, changes are minimal, consisting of matters of interpretation and guidance, not requirement or specification. The role of human rights has been clarified by reference to the United Nations Guiding Principles on Business and Human Rights. Greater emphasis has been put on the absolute priority to ensure flag State requirements are identified and met. The different concepts of "threat assessment" and "risk" have been clarified. The phrase "interested parties" has been replaced by "stakeholders" for textual consistency, and "reasonable and proportionate" has been replaced with "reasonable and necessary".[84]

Box 5.1. The current status of the ISO 28000 series of standards

Standards published:

- **ISO 28000:2007** – "Specification for security management systems for the supply chain". This standard provides the overall "umbrella" standard. It is a generic, risk-based, certifiable standard for all organizations, all disruptions and all sectors. It is widely in use and constitutes a stepping stone to the AEO and C–TPAT certifications.

- **ISO 28001:2007** – "Security management systems for the supply chain – Best practices for implementing supply chain security, assessments and plans". This standard is designed to assist the industry to meet the requirements for AEO status.

- **ISO 28002:2011** – "Security management systems for the supply chain – Development of resilience in the supply chain – Requirements with guidance for use". This standard provides additional focus on resilience, and emphasizes the need for an ongoing, interactive process to prevent, respond to and assure continuation of an organization's core operations after a major disruptive event.

- **ISO 28003:2007** – "Security management systems for the supply chain – Requirements for bodies providing audit and certification of supply chain security management systems". This standard provides guidance for accreditation and certification bodies.

- **ISO 28004-1:2007** – "Security management systems for the supply chain – Guidelines for the implementation of ISO 28000 – Part 1: General principles". This standard provides generic advice on the application of ISO 28000:2007. It explains the underlying principles of ISO 28000 and describes the intent, typical inputs, processes and typical outputs for each requirement of ISO 28000. The objective is to aid the understanding and implementation of ISO 28000. ISO 28004-1:2007 does not create additional requirements to those specified in ISO 28000, nor does it prescribe mandatory approaches to the implementation of ISO 28000.

- **ISO/PAS 28004-2:2014** – "Security management systems for the supply chain – Guidelines for the implementation of ISO 28000 – Part 2: Guidelines for adopting ISO 28000 for use in medium and small seaport operations". This standard provides guidance to medium-sized and small ports that wish to adopt ISO 28000. It identifies supply chain risk and threat scenarios, procedures for conducting risk/threat assessments, and evaluation criteria for measuring conformance and effectiveness of the documented security plans in accordance with ISO 28000 and ISO 28004 implementation guidelines.

- **ISO/PAS 28004-3:2014** – "Security management systems for the supply chain – Guidelines for the implementation of ISO 28000 – Part 3: Additional specific guidance for adopting ISO 28000 for use by medium and small businesses (other than marine ports)". This standard has been developed to supplement ISO 28004-1 by providing additional guidance to small and medium-sized businesses (other than marine ports) that wish to adopt ISO 28000. The additional guidance in ISO/PAS 28004-3:2012, while amplifying the general guidance provided in the main body of ISO 28004-1, does not conflict with the general guidance, nor does it amend ISO 28000.

- **ISO/PAS 28004-4:2014** – "Security management systems for the supply chain – Guidelines for the implementation of ISO 28000 – Part 4: Additional specific guidance on implementing ISO 28000 if compliance with ISO 28001 is a management objective". This standard provides additional guidance for organizations adopting ISO 28000 that also wish to incorporate the best practices identified in ISO 28001 as a management objective in their international supply chains.

- **ISO 28005-1:2013** – "Security management systems for the supply chain – Electronic port clearance (EPC) – Part 1: Message structures". This standard provides for computer-to-computer data transmission.

- **ISO 28005-2:2011** – "Security management systems for the supply chain – Electronic port clearance (EPC) – Part 2: Core data elements". This standard contains technical specifications that facilitate efficient exchange of electronic information between ships and shore for coastal transit or port calls, as well as definitions of core data elements that cover all requirements for ship-to-shore and shore-to-ship reporting as defined in the ISPS Code, the IMO Facilitation Committee Convention and relevant IMO resolutions.

- **ISO/PAS 28007-1:2015** – "Ships and marine technology – Guidelines for private maritime security companies (PMSC) providing privately contracted armed security personnel on board ships (and pro forma contract) – Part 1: General". This standard provides guidelines containing additional sector-specific recommendations, which companies (organizations) that comply with ISO 28000 can implement to demonstrate that they provide privately contracted armed security personnel on board ships.

> **Box 5.1. The current status of the ISO 28000 series of standards** *(continued)*
>
> - **ISO 20858:2007** – "Ships and marine technology – Maritime port facility security assessments and security plan development". This standard establishes a framework to assist marine port facilities in specifying the competence of personnel to conduct a marine port facility security assessment and to develop a security plan as required by the ISPS Code. In addition, it establishes certain documentation requirements designed to ensure that the process used in performing the duties described above is recorded in a manner that permits independent verification by a qualified and authorized agency. It is not an objective of ISO 20858:2007 to set requirements for a contracting Government or designated authority in designating a recognized security organization, or to impose the use of an outside service provider or other third parties to perform the marine port facility security assessment or security plan if the port facility personnel possess the expertise outlined in the specification. Ship operators may be informed that marine port facilities that use this document meet an industry-determined level of compliance with the ISPS Code. ISO 20858:2007 does not address the requirements of the ISPS Code relative to port infrastructure that fall outside the security perimeter of a marine port facility that might affect the security of the facility–ship interface. Governments have a duty to protect their populations and infrastructures from marine incidents occurring outside their marine port facilities. These duties are outside the scope of ISO 20858:2007.
>
> **Standards under development:**
>
> - ISO 28006 – "Security management systems for the supply chain – Security management of [roll-on roll-off] RO-RO passenger ferries". This standard will include best practices for the application of security measures.
>
> *Note:* For more information, including on the procedure for preparing international standards at ISO, see www.iso.org.

2. Combating maritime piracy and armed robbery

At a basic level, maritime piracy is a maritime transport issue that directly affects ships, ports, terminals, cargo and seafarers. However, as piracy activities evolve and become more sophisticated, the problem becomes a multifaceted and complex transnational security challenge that threatens lives, livelihoods and global welfare. As highlighted in some detail in a recent two-part report on maritime piracy prepared by UNCTAD, piracy has broad repercussions, including for humanitarian aid, supply chains, global production processes, trade, energy security, fisheries, marine resources, the environment and political stability (UNCTAD 2014b, 2014c). The resulting adverse and potentially destabilizing effects entail far-reaching implications for all countries, whether they are coastal or landlocked, developed or developing. Addressing the challenge of piracy in an effective manner requires strong cooperation at the political, economic, legal, diplomatic and military levels, as well as collaboration between diverse public and private sector stakeholders across regions.

At IMO, MSC at its ninety-fourth session (17–21 November 2014) welcomed the continuing positive developments in the suppression of piracy and armed robbery in the waters off the coast of Somalia and the wider western Indian Ocean, but remained concerned about the seafarers still being held hostage. It also noted the downward trend of attacks in the Gulf of Guinea, indicating that international, regional and national efforts were beginning to take effect, and reiterated the importance of reporting incidents by flag States and industry organizations.[85]

MSC noted the work of the Maritime Trade Information Sharing Centre,[86] now operational on a trial basis with over 500 ships per month reporting to it. The work of the Maritime Trade Information Sharing Centre is complementary to that of the Interregional Coordination Centre in Yaoundé. The latter provides for cooperation, coordination and communication in the implementation[87] of the Code of Conduct concerning the repression of piracy, armed robbery against ships and illicit maritime activity in West and Central Africa[88] at the strategic level, while the Maritime Trade Information Sharing Centre handles civilian information exchange and maritime situational awareness aspects.[89] MSC expressed its appreciation for the contributions received for the IMO West and Central Africa Maritime Security Trust Fund,[90] and called on member States to further support the implementation of IMO projects on maritime security for West and Central Africa by financially contributing to the Trust Fund.

With respect to piracy off the coast of Somalia, MSC noted United Nations Security Council resolution 2184 on the situation in Somalia, adopted on 12 November

2014, which, among others, recognized the contribution of IMO and renewed its call upon States to deploy naval vessels to the area, and underlined the primary responsibility of Somali authorities in the fight against piracy and armed robbery off the country's coast (United Nations, 2014a). MSC also welcomed the fact that the European Union Naval Force and North Atlantic Treaty Organization mandates had been extended to the end of 2016, and reiterated the importance of continuing to implement diligently the IMO guidance and best management practices.[91]

As regards the situation of piracy and armed robbery against ships in Asia for the period January to June 2014, MSC noted a document providing an update on the activities carried out by the Information Sharing Centre of the Regional Cooperation Agreement on Combating Piracy and Armed Robbery against Ships in Asia, and including the action taken by some of its members with respect to those found to be responsible for fuel siphoning cases that had been widely reported in the media (IMO, 2014f).[92]

MSC also noted United Nations Security Council resolution 2182 on the situation in Somalia and Eritrea, adopted on 24 October 2014, highlighting the need to prevent unauthorized deliveries of weapons and military equipment to Somalia and to prevent the direct or indirect export of charcoal from Somalia (United Nations, 2014b). Some of its provisions, particularly operative paragraph 10 in relation to weapons on board vessels engaged in commercial activity in Somali ports, and operative paragraphs 11 to 22 referring to the maritime interdiction of charcoal and arms, may have an impact on the shipping industry. Implications may also arise from paragraphs 15 and 16, dealing specifically with inspections by member States, acting nationally or through voluntary multinational naval partnerships, of merchant ships that they have reasonable grounds to believe are carrying charcoal or weapons in violation of the ban and/or embargo.

MSC at its ninety-fifth session approved:

- "Recommendations to Governments for preventing and suppressing piracy and armed robbery against ships", which incorporates a provision on the establishment of a national point of contact for communication of information on piracy and armed robbery to IMO (IMO, 2015h);

- "Best management practices for protection against Somalia-based piracy" (IMO, 2015i);

- "Revised interim recommendations for flag States regarding the use of privately contracted armed security personnel on board ships in the high risk area", which includes amendments related to certification of private maritime security companies to address publication of ISO 28007 (IMO, 2015j).

The Legal Committee at its 102nd session considered a document by the secretariat (IMO, 2015k) reporting on the outcome of discussions by members of the Kampala Process[93] at a meeting led by IMO, with the support of EUCAP Nestor[94] and the United Nations Office on Drugs and Crime, held in Addis Ababa in September 2014. The Committee was also informed[95] of the current status of the secretariat's counter-piracy initiatives.[96]

3. Seafarers' issues

Shipping and related activities are expected to continue to provide important opportunities for employment in developing countries, thus contributing to achieving sustainable development goals. According to ILO estimates, over 1.5 million people around the world are employed as seafarers, the vast majority of whom come from developing countries.[97] Protecting their welfare and establishing internationally agreed standards, including on their working conditions and necessary training, is critical, not only for the seafarers themselves, but also for the ability of the global shipping industry to operate ships safely and in an environmentally responsible manner.

The most important and comprehensive international instrument negotiated at ILO, the MLC 2006, which consolidates and updates more than 68 international labour standards relating to seafarers, and sets out their responsibilities and rights with regard to labour and social matters in the maritime sector, entered into force on 20 August 2013. It currently has 65 member States, representing over 80 per cent of the world's global shipping tonnage, and is considered as the fourth pillar of the global maritime regulatory regime. Therefore, the review of the implementation of the MLC on a regular basis, as well as consultations regarding any necessary updates to it, are very important. Worth noting are the 2014 amendments to the MLC aimed at ensuring that adequate financial security is provided by flag States to cover the costs of abandonment of seafarers as well as claims for death and long-term disability due to occupational injury and hazards, thus providing relief to seafarers and their families and improving the quality of shipping overall. These amendments, which were summarized in the

Review of Maritime Transport 2014, were approved by the International Labour Conference held in June 2014 (UNCTAD, 2014a, pages 89–90).

(a) International Labour Organization Convention No. 185 on Seafarers' Identity Documents (Revised) 2003

Convention No.185 specifically relates to the issuance and recognition of the seafarers' identity document (SID) which facilitates the temporary admission of seafarers to foreign territory, for the purposes of their well-being while in port, accessing onshore welfare facilities or taking shore leave, and for transit through a country related to the operation of ships.[98] A SID can only be issued and verified by a seafarer's country of nationality. Although SIDs are not considered travel documents per se (such as, for example, passports or visas), their issuance may be subject to the same conditions as prescribed by national laws for travel documents.

Convention No. 185, adopted in June 2003 to replace the earlier Convention No. 108, included innovations that related to the introduction of modern security features at the time, for the new SID and its biometric features (fingerprint template and photograph), as well as features facilitating verification of the SID (uniformity and machine readability). Convention No. 185 also contains minimum requirements for SID issuance processes and procedures, including quality control, national databases and national focal points to provide information to border authorities.

Although the Convention entered into force in February 2005, only 30 out of 185 ILO member States have ratified it or provisionally applied it to date, and this number includes few port States. Thus, those countries that have made considerable investments to properly implement this Convention can count on only a few countries to recognize the SIDs issued under it. Also, only a few countries that have ratified the Convention are in a position to actually issue SIDs conforming to it. Implementation efforts are mainly hampered by the fact that the fingerprint technology and biometric features required in annex I of Convention No. 185 are already considered to be out of date, and are not used by the border authorities of many of the countries concerned. Instead, since 2003, many of these countries have been using the International Civil Aviation Organization standards for travel documents, which are exclusively based on the facial image in a contactless chip as the biometric feature, rather than a fingerprint template in a two-dimensional barcode.

After careful consideration of these matters, participants at the Tripartite Meeting of Experts concluded that the only feasible way forward would be for the 2016 International Labour Conference to amend annex I to Convention No. 185, and as necessary other annexes, to align the biometric requirements under this Convention with those of the International Civil Aviation Organization that are universally followed for travel and similar documents. However, a suitable transition period would be allowed for countries that are already implementing Convention No. 185.[99]

(b) Fair treatment of seafarers in the event of a maritime accident

The Legal Committee at its 102nd session considered the outcome of a survey concerning implementation of the 2006 Guidelines on the Fair Treatment of Seafarers in the Event of a Maritime Accident, and a further analysis of the responses to this survey (IMO, 2015l).[100] The survey indicated the following:

- Thirteen member States (29 per cent of the respondents) stated that their existing laws already adequately protect the human and other legal rights of seafarers contained in the guidelines and that, therefore, there was no need for the guidelines to be passed into their existing laws;

- Seventeen member States (38 per cent of the respondents) had passed the guidelines, either in whole or in part, into their national laws, either explicitly or implicitly;

- Fifteen member States (33 per cent of the respondents) requested assistance in the form of information regarding the meaning of the guidelines and/or model legislation by IMO for the purpose of giving effect to the guidelines.

The Committee concluded that (IMO, 2015g, pages 6–7):

- [T]his was an important issue for seafarers and should consequently be placed on the work programme of the Legal Committee;

- [T]he Committee should consider guidance on the implementation of the Guidelines, in particular for developing countries;

- [T]echnical support and assistance should be provided by [the Technical Cooperation Committee] TCC in order to facilitate the wide implementation of the Guidelines to improve the conditions for seafarers, taking into account human rights issues;

- [W]ork needed to be done towards the progressive removal of legislation targeting seafarers and imposing criminal sanctions on them;
- [I]t would be useful for States already giving effect to the Guidelines to provide translated copies of their laws to assist other States with their implementation efforts; and some States informed the Committee that they were ready to share their national legislation giving effect to the Guidelines;
- [W]ith regard to the compilation of statistics, it was also relevant to receive feedback from ports;
- States were urged to provide their embassies with the names of persons whom seafarers could contact to report violations of the Guidelines;
- [S]eafarers should be given greater training and awareness of their rights.

The Committee also noted with gratitude that the industry was prepared to contribute financially towards this work.

Key developments in summary

During the year under review, continued enhancements were made to regulatory measures in the field of maritime and supply chain security and their implementation. Developments included the issuance of a new version of the WCO SAFE Framework in June 2015, which includes a new pillar 3 aiming to foster closer cooperation between customs administrations and other government agencies involved in the international trade supply chain. Other areas of progress included the implementation of AEO programmes and an increasing number of bilateral MRAs that will, in due course, form the basis for the recognition of AEOs at a multilateral level. As regards suppression of piracy and armed robbery, positive developments were noted in the waters off the coast of Somalia and the wider western Indian Ocean. However, concern remained about the seafarers still being held hostage. A downward trend of attacks in the Gulf of Guinea was also observed, indicating that international, regional and national efforts were beginning to take effect. Progress was also made at ILO and IMO regarding issues related to seafarers' fair treatment in the event of a maritime accident as well as to the issuance and recognition of SIDs.

D. STATUS OF CONVENTIONS

A number of international conventions in the field of maritime transport were prepared or adopted under the auspices of UNCTAD. Table 5.1 provides information on the status of ratification of each of these conventions as at 30 June 2015.

CHAPTER 5: LEGAL ISSUES AND REGULATORY DEVELOPMENTS

Table 5.1. Contracting States Parties to selected international conventions on maritime transport as at 30 June 2015

Title of convention	Date of entry into force or conditions for entry into force	Contracting States
United Nations Convention on a Code of Conduct for Liner Conferences, 1974	Entered into force 6 October 1983	Algeria, Bangladesh, Barbados, Belgium, Benin, Burkina Faso, Burundi, Cameroon, Cabo Verde, Central African Republic, Chile, China, Congo, Costa Rica, Côte d'Ivoire, Cuba, Czech Republic, Democratic Republic of the Congo, Egypt, Ethiopia, Finland, France, Gabon, Gambia, Ghana, Guatemala, Guinea, Guyana, Honduras, India, Indonesia, Iraq, Italy, Jamaica, Jordan, Kenya, Kuwait, Lebanon, Liberia, Madagascar, Malaysia, Mali, Mauritania, Mauritius, Mexico, Montenegro, Morocco, Mozambique, Niger, Nigeria, Norway, Pakistan, Peru, Philippines, Portugal, Qatar, Republic of Korea, Romania, Russian Federation, Saudi Arabia, Senegal, Serbia, Sierra Leone, Slovakia, Somalia, Spain, Sri Lanka, Sudan, Sweden, Togo, Trinidad and Tobago, Tunisia, United Republic of Tanzania, Uruguay, Venezuela (Bolivarian Republic of), Zambia (76)
United Nations Convention on the Carriage of Goods by Sea, 1978 (Hamburg Rules)	Entered into force 1 November 1992	Albania, Austria, Barbados, Botswana, Burkina Faso, Burundi, Cameroon, Chile, Czech Republic, Dominican Republic, Egypt, Gambia, Georgia, Guinea, Hungary, Jordan, Kazakhstan, Kenya, Lebanon, Lesotho, Liberia, Malawi, Morocco, Nigeria, Paraguay, Romania, Saint Vincent and the Grenadines, Senegal, Sierra Leone, Syrian Arab Republic, Tunisia, Uganda, United Republic of Tanzania, Zambia (34)
International Convention on Maritime Liens and Mortgages, 1993	Entered into force 5 September 2004	Albania, Benin, Congo, Ecuador, Estonia, Lithuania, Monaco, Nigeria, Peru, Russian Federation, Spain, Saint Kitts and Nevis, Saint Vincent and the Grenadines, Serbia, Syrian Arab Republic, Tunisia, Ukraine, Vanuatu (18)
United Nations Convention on International Multimodal Transport of Goods, 1980	Not yet in force – requires 30 contracting Parties	Burundi, Chile, Georgia, Lebanon, Liberia, Malawi, Mexico, Morocco, Rwanda, Senegal, Zambia (11)
United Nations Convention on Conditions for Registration of Ships, 1986	Not yet in force – requires 40 contracting Parties with at least 25 per cent of the world's tonnage as per annex III to the Convention	Albania, Bulgaria, Côte d'Ivoire, Egypt, Georgia, Ghana, Haiti, Hungary, Iraq, Liberia, Libya, Mexico, Morocco, Oman, Syrian Arab Republic (15)
International Convention on Arrest of Ships, 1999	Entered into force 14 September 2011	Albania, Algeria, Benin, Bulgaria, Congo, Ecuador, Estonia, Latvia, Liberia, Spain, Syrian Arab Republic (11)

Note: For official status information, see http://treaties.un.org (accessed 24 September 2015).

E. TRADE FACILITATION AND SUSTAINABLE DEVELOPMENT

International shipping is also affected by the facilitation of maritime trade, that is, the import and export procedures and documentation requirements in seaports. Trade facilitation aims at simplifying administrative procedures and making them transparent and less time consuming and cumbersome for users involved in foreign trade operations. This will benefit concerned public sector agencies and traders, while improving transparency and governance. In this context, trade facilitation reforms are increasingly incorporated into broader policy areas that are relevant for achieving the SDGs. Beyond their relevance for trade competitiveness, most specific trade facilitation reforms also have a direct impact on a number of sustainable development targets.

Trade facilitation reforms and development mutually benefit each other in various ways (see Kituyi, 2013, 2014). The most frequently mentioned linkage is the positive impact that trade facilitation has on the competitiveness of developing countries and their participation in global trade and value chains (WTO, 2015a). Apart from this well-known impact that trade facilitation reforms have on trade, there exist important additional linkages with a country's development.

The entry into force and implementation of the TFA contributes towards "a universal, rules-based, open, non-discriminatory and equitable multilateral trading system under the WTO" (SDG target 17.10). The technical assistance and capacity-building to be provided under section II of the TFA can help to "increase Aid for Trade support for developing countries, particularly LDCs, including through the Enhanced Integrated

Table 5.2.	Examples of articles of the TFA that may benefit from and help to achieve SDGs
Articles of the WTO TFA	**Selected extracts from SDGs**
Article 1: Publication and availability of information	"public access to information" (16.10)
Article 2: Opportunity to comment, information before entry into force and consultations	"responsive, inclusive, participatory and representative decision-making at all levels" (16.7)
Article 3: Advance rulings	"develop effective, accountable and transparent institutions at all levels" (16.6)
Article 4: Procedures for appeal or review	"rule of law at the national and international levels, and ensure equal access to justice for all" (16.3)
Article 5: Other measures to enhance impartiality, non-discrimination and transparency	"access to information and communications technology and strive to provide universal and affordable access to Internet in LDCs" (9.c)
Article 6: Disciplines on fees and charges imposed on or in connection with importation and exportation and penalties	"reduce corruption and bribery" (16.5)
Article 7: Release and clearance of goods	"enhance the use of enabling technologies in particular information and communications technology" (17.8)
Article 8: Border agency cooperation	"higher levels of economic productivity through diversification, technological upgrading and innovation" (8.2)
Article 9: Movement of goods intended for import under customs control	"capacity of domestic financial institutions" (8.10)
Article 10: Formalities connected with importation, exportation and transit	"higher levels of economic productivity through diversification, technological upgrading and innovation" (8.2)
Article 11: Freedom of transit	"regional and trans-border infrastructure" (9.1)
Article 12: Customs cooperation	"strengthen relevant national institutions, including through international cooperation, for building capacities at all levels, in particular in developing countries, for preventing violence and combating terrorism and crime" (16.a)
Article 23.2: National Committee on Trade Facilitation	"effective public, public–private and civil society partnerships" (17.17)

Source: Open Working Group proposal for sustainable development goals, available at https://sustainabledevelopment.un.org/focussdgs.html (accessed 25 June 2015).

Framework" (SDG target 8.a) and can "enhance international support for implementing effective and targeted capacity-building in developing countries to support national plans to implement all sustainable development goals, including through North–South, South–South and triangular cooperation" (SDG target 17.9).

Many of the specific trade facilitation measures that are included in the TFA also have a direct linkage to different aspects of development. Table 5.2 provides a list of articles included in the TFA and links them to selected SDGs and targets.

Article 1 of the TFA, for example, covers the publication and availability of information on import, export and transit procedures; a country that complies with article 1 of the TFA will thus be closer to achieving the SDG target 16.10, which, inter alia, aims at ensuring "public access to information". Another example is article 5 of the TFA, which, inter alia, requires Governments to publish certain announcements in a non-discriminatory and easily accessible manner; this is more easily achieved if traders have "access to Internet", as stipulated in SDG target 9.c. Article 6 of the TFA includes the requirement to avoid "conflicts of interest in the assessment and collection of penalties and duties", which can help to "reduce corruption and bribery" covered by SDG target 16.5. A further example of possible linkages between the TFA and the SDGs is TFA article 11 on freedom of transit, which complements "regional and trans-border infrastructure" covered by SDG target 9.1.

For the effective implementation of the TFA, WTO members are required to "establish and/or maintain a national committee on trade facilitation or designate an existing mechanism to facilitate both domestic coordination and implementation of the provisions of [the TFA]". Such a mechanism is crucial for ensuring the political buy-in of the relevant stakeholders from the public and private sectors, including users and providers of trade-supporting services (UNCTAD, 2014d). It also responds to the SDG target 17.17 to "encourage and promote effective public, public–private and civil society partnerships, building on the experience and resourcing strategies of partnerships".

CHAPTER 5: LEGAL ISSUES AND REGULATORY DEVELOPMENTS

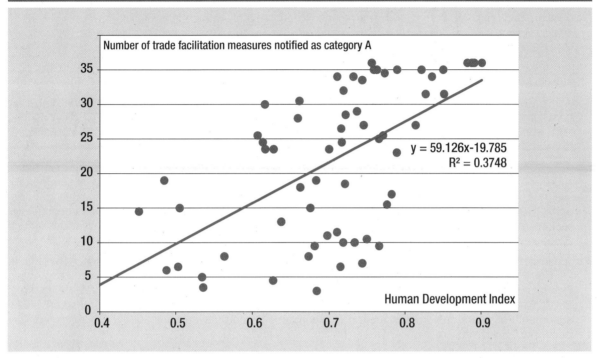

Figure 5.1. The Human Development Index (HDI) and the number of trade facilitation measures notified as category A

Source: UNCTAD secretariat, based on individual notifications published on the WTO website http://www.wto.org/english/tratop_e/tradfa_e/tradfa_e.htm#notifications (accessed 24 September 2015). The HDI is sourced from UNDP, available at http://hdr.undp.org/en/content/human-development-index-hdi (accessed 24 September 2015).

In addition to the specific SDGs mentioned in table 5.2, there are several cross-cutting SDGs that benefit from and help to implement trade facilitation reforms. "[E]qual access for all women and men to affordable quality technical, vocational and tertiary education, including university" (4.3), for example, will help strengthen the capacities of traders and service providers to make use of the latest technologies and methods utilized by customs administrations and other border agencies. In general, many trade facilitation measures help the informal sector to better participate in formal foreign trade, thus supporting SDG target 8.3 on the "formalization and growth of micro-, small- and medium-sized enterprises".

Since early 2014, WTO members have started to notify their "category A" trade facilitation measures to the WTO. "Category A contains provisions that a developing country Member or a least-developed country Member designates for implementation upon entry into force of this Agreement, or in the case of a least-developed country Member within one year after entry into force" (WTO, 2014). By 30 July 2015, a total of 67 developing countries had notified their category A provisions to the WTO secretariat (WTO, 2015b).

An analysis of the number of category A measures notified per country suggests that a close correlation exists between different indicators for development and the implementation of trade facilitation reforms. While a statistical correlation does not in itself say anything about causalities, the data suggest that the possible linkages listed in table 5.2 are supported by empirical evidence.[101] For example, the coefficient of determination R^2 between the HDI and the number of measures notified as category A is around 0.37, suggesting that about 37 per cent of the variation in the number of category A notifications per country is statistically explained by the country's HDI (figure 5.1).

Interestingly, the implementation of trade facilitation measures as reflected in the category A notifications is statistically less correlated with a country's trade than with its level of development, as measured by the GDP per capita or the HDI. Put differently, the data from the category A notifications suggest that the likelihood that a developing country will implement trade facilitation reforms has more to do with its capacity and human and institutional development than with its level of foreign trade. Capacity development will thus continue to be key for the advancement of the TFA on the ground.

REFERENCES

Bergerson SG (2008). Arctic meltdown, the economic and security implications of global warming. *Foreign Affairs*. March/April.

Det Norske Veritas (2011). Polar Code Hazard Identification Workshop report for IMO. 25 October. Available at http://www.imo.org/en/MediaCentre/HotTopics/polar/Documents/INF-3%20annex%20Polar%20Code%20 Workshop%20Report%2025Oct11.pdf (accessed 9 September 2015).

Economic Commission for Europe (2013). *Climate Change Impacts and Adaptation for International Transport Networks.* Expert group report. Inland Transport Committee. United Nations publication. ECE/TRANS/238. New York and Geneva. Available at http://www.unece.org/fileadmin/DAM/trans/main/wp5/publications/climate_change_2014.pdf (accessed 6 July 2015).

European Commission (2014a). Communication from the Commission to the European Parliament, the Council and the European Economic and Social Committee on the European Union Strategy and Action Plan for customs risk management: Tackling risks, strengthening supply-chain security and facilitating trade. COM(2014) 527 final. Brussels. Available at http://ec.europa.eu/taxation_customs/resources/documents/customs/customs_controls/risk_management/customs_eu/com_2014_527_en.PDF (accessed 12 June 2015).

European Commission (2014b). Annex to the Communication from the Commission to the European Parliament, the Council and the European Economic and Social Committee on the European Union Strategy and Action Plan for customs risk management: Tackling risks, strengthening supply-chain security and facilitating trade. COM(2014) 527 final. Annex I. Brussels. Available at http://ec.europa.eu/taxation_customs/resources/documents/customs/customs_controls/risk_management/customs_eu/com_2014_527_annex_en.PDF (accessed 12 June 2015).

IHS Maritime Technlogy (2014). Emissions control. November.

IMO (2002). Guidelines for ships operating in Arctic ice-covered waters. MSC/Circ.1056. MEPC/Circ.399. London.

IMO (2009). Guidelines for ships operating in polar waters. Resolution A.1024(26). London.

IMO (2010). Report of the Marine Environment Protection Committee at its sixty-first session. MEPC 61/24. London.

IMO (2011). Report of the Marine Environment Protection Committee at its sixty-second session. MEPC 62/24. London.

IMO (2013). Sulphur monitoring programme for fuel oils for 2012. MEPC 65/4/9. London.

IMO (2014a). Third IMO GHG study 2014 – Final report. MEPC 67/INF.3. London.

IMO (2014b). Report of the Marine Environment Protection Committee on its sixty-seventh session. MEPC 67/20. London.

IMO (2014c). Sulphur monitoring for 2013. MEPC 67/4. London.

IMO (2014d). Amendments to MARPOL. Circular letter No. 3495 of 30 October 2014. London.

IMO (2014e). Report of the Ballast Water Review Group. MEPC 68/WP.8. London.

IMO (2014f). Progress report of the Regional Cooperation Agreement on Combating Piracy and Armed Robbery against Ships in Asia (ReCAAP) Information Sharing Centre (ISC). Submitted by the ReCAAP-ISC. MSC 94/INF.7. London.

IMO (2014g). Developments since MSC 93. MSC 94/14. London.

IMO (2014h). Report of the Maritime Safety Committee on the work of its ninety-fourth session. MSC 94/21. London.

IMO (2015a). Report of the Marine Environment Protection Committee on its sixty-eighth session. MEPC 68/21. London.

IMO (2015b). Outcomes of the United Nations Climate Change Conferences held in Lima in December 2014 and in Geneva in February 2015. MEPC 68/5. London.

IMO (2015c). Guidance on the application of regulation 13 of MARPOL annex VI tier III requirements to dual fuel and gas-fuelled engines. MEPC.1/Circ.854. London.

IMO (2015d). Sulphur monitoring for 2014. MEPC 68/3/2. London.

IMO (2015e). Report [of the Sub-committee on Pollution Prevention and Response] to the Marine Environment Protection Committee. PPR 2/21/Add.1. London.

IMO (2015f). Progress report on the study on the implementation of the ballast water performance standard described in regulation D-2 of the BWM Convention. MEPC 68/2/11. London.

IMO (2015g). Report of the Legal Committee on the work of its one hundred and second session. LEG 102/12. London.

IMO (2015h). Recommendations to Governments for preventing and suppressing piracy and armed robbery against ships. MSC.1/Circ.1333/Rev.1. London.

IMO (2015i). Best management practices for protection against Somalia-based piracy. MSC.1/Circ.1506. London.

IMO (2015j). Revised interim recommendations for flag States regarding the use of privately contracted armed security personnel on board ships in the high risk area. MSC.1/Circ.1406/Rev.3. London.

IMO (2015k). Piracy. LEG 102/5. London.

IMO (2015l). Analysis of the questionnaire on the implementation of the 2006 guidelines on fair treatment of seafarers in the event of a maritime accident. Submitted by the International Transport Workers' Federation, the International Federation of Shipmasters' Associations, the Comité Maritime International and InterManager. LEG 102/4. London.

Kituyi M (2013). Trade facilitation: Trade competitiveness and the development dimension. International Trade Centre – Trade Forum. 1 December. Available at http://www.tradeforum.org/article/Trade-facilitation-Trade-competitiveness-and-the-development-dimension/ (accessed 9 September 2015).

Kituyi M (2014). Cutting red tape in trade supports development. *Huffington Post*. 2 December. Available at http://www.huffingtonpost.com/mukhisa-kituyi/cutting-red-tape-in-trade_b_6248562.html (accessed 9 September 2015).

The Ship Supplier (2014). Tier III rules will become cost saver. Issue 61.

UNCTAD (2004). Container security: Major initiatives and related international developments. UNCTAD/SDTE/TLB/2004/1. Available at http://unctad.org/en/Docs/sdtetlb20041_en.pdf (accessed 12 June 2015).

UNCTAD (2007). Maritime security: ISPS code implementation, costs and related financing. UNCTAD/SDTE/TLB/2007. Available at http://unctad.org/en/Docs/sdtetlb20071_en.pdf (accessed 24 September 2015).

UNCTAD (2009). Multi-Year Expert Meeting on Transport and Trade Facilitation: Maritime Transport and the Climate Change Challenge. Summary of proceedings. UNCTAD/DTL/TLB/2009/1. Geneva. Available at http://unctad.org/en/Docs/dtltlb20091_en.pdf (accessed 17 September 2015).

UNCTAD (2011a). *Review of Maritime Transport 2011*. United Nations publication. Sales No. E.11.II.D.4. New York and Geneva. Available at (accessed 30 September 2015).

UNCTAD (2011b). The 2004 Ballast Water Management Convention – with international acceptance growing, the Convention may soon enter into force. Transport Newsletter No. 50, second quarter: 8–12. Available at http://unctad.org/en/Docs/webdtltlb20113_en.pdf (accessed 30 September 2015).

UNCTAD (2012a). *Liability and Compensation for Ship-source Oil Pollution: An Overview of the International Legal Framework for Oil Pollution Damage from Tankers.* Studies in transport law and policy 2012 No. 1. United Nations publication. UNCTAD/DTL/TLB/2011/4. New York and Geneva. Available at http://unctad.org/en/PublicationsLibrary/dtltlb20114_en.pdf (accessed 24 September 2015).

UNCTAD (2012b). *Review of Maritime Transport 2012*. United Nations publication. Sales no. E.12.II.D.17. New York and Geneva. Available at http://unctad.org/en/PublicationsLibrary/rmt2012_en.pdf (accessed 12 June 2015).

UNCTAD (2013). *Review of Maritime Transport 2013*. United Nations publication. Sales No. E.13.II.D.9. New York and Geneva. Available at http://unctad.org/en/PublicationsLibrary/rmt2013_en.pdf (accessed 12 June 2015).

UNCTAD (2014a). *Review of Maritime Transport 2014*. United Nations publication. Sales No. E.14.II.D.5. New York and Geneva. Available at http://unctad.org/en/PublicationsLibrary/rmt2014_en.pdf (accessed 24 July 2015).

UNCTAD (2014b). *Maritime Piracy. Part I: An Overview of Trends, Costs and Trade-related Implications.* United Nations publication. UNCTAD/DTL/TLB/2013/1. New York and Geneva. Available at http://unctad.org/en/PublicationsLibrary/dtltlb2013d1_en.pdf (accessed 24 June 2015).

UNCTAD (2014c). *Maritime Piracy. Part II: An Overview of the International Legal Framework and of Multilateral Cooperation to Combat Piracy.* United Nations publication. UNCTAD/DTL/TLB/2013/3. New York and Geneva. Available at http://unctad.org/en/PublicationsLibrary/dtltlb2013d3_en.pdf (accessed 24 June 2015).

UNCTAD (2014d). *National Trade Facilitation Bodies in the World.* United Nations publication. UNCTAD/DTL/TLB/2014/1. New York and Geneva. Available at http://unctad.org/en/Pages/DTL/Trade-Logistics-Branch.aspx (accessed 17 September 2015).

United Nations (2014a). United Nations Security Council resolution 2184. 12 November. S/RES/2184. New York.

United Nations (2014b). United Nations Security Council resolution 2182. 24 October. S/RES/2182. New York.

United Nations Environment Programme (2015). Message by the United Nations Secretary-General, Ban Ki-moon on World Oceans Day. 8 June. Available at http://www.unep.org/newscentre/Default.aspx?DocumentID=26827&ArticleID=35178&l=en (accessed 9 September 2015).

WCO (2011). The customs supply chain security paradigm and 9/11: Ten years on and beyond. WCO research paper No.18. September. Available at http://www.wcoomd.org/~/media/WCO/Public/Global/PDF/Topics/Research/Research%20Paper%20Series/18_CSCSP_911.ashx?db=web (accessed 28 September 2015).

WCO (2012). *SAFE Framework of Standards to Secure and Facilitate Global Trade.* June. Available at http://www.wcoomd.org/en/topics/facilitation/instrument-and-tools/tools/~/media/55F00628A9F94827B58ECA90C0F84F7F.ashx (accessed 12 June 2014).

WCO (2015a). *SAFE Framework of Standards to Secure and Facilitate Global Trade.* June. Available at http://www.wcoomd.org/en/topics/facilitation/instrument-and-tools/tools/~/media/2B9F7D493314432BA42BC8498D3B73CB.ashx (accessed 28 September 2015).

WCO (2015b). *Compendium of Authorized Economic Operator Programmes.* 2015 edition. Available at http://www.wcoomd.org/en/topics/facilitation/instrument-and-tools/tools/~/media/3109C877081E4071B4E2C938317CBA9C.ashx (accessed 28 September 2015).

Wilson KJ, Falkingham J, Melling H and De Abreu R (2004). Shipping in the Canadian Arctic: Other possible climate change scenarios. Canadian Ice Service and the Institute of Ocean Sciences. Victoria.

WTO (2014). Agreement on Trade Facilitation. Article 14: Categories of provisions. WT/L/931. 15 July. Available at http://www.wto.org/english/news_e/news14_e/sum_gc_jul14_e.htm (accessed 9 September 2015).

WTO (2015a). Doha Development Agenda. Available at http://www.wto.org/english/thewto_e/coher_e/mdg_e/dda_e.htm (accessed 17 September 2015).

WTO (2015b). Notifications of category A commitments, available at https://www.wto.org/english/tratop_e/tradfa_e/tradfa_e.htm#notifications (accessed 28 September 2015).

CHAPTER 5: LEGAL ISSUES AND REGULATORY DEVELOPMENTS

ENDNOTES

1. A new chapter XIV, "Safety measures for ships operating in polar waters".
2. According to the tacit acceptance procedure, amendments enter into force by default unless objections are filed by a certain number of States.
3. Relevant in this context is a recent resolution by the United Nations General Assembly (A/69/L.65) deciding to develop an internationally legally binding instrument under UNCLOS on the conservation and sustainable use of marine biological diversity of areas beyond national jurisdiction.
4. See the Antarctic and Southern Ocean Coalition (a coalition of over 30 non-governmental organizations interested in Antarctic environmental protection and conservation) press release, available at http://www.asoc.org/explore/latest-news/1364-press-release-polar-code-too-weak-to-properly-protect-polar-environments-from-increased-shipping-activity (accessed 9 September 2015).
5. See the IMO press release available at http://www.imo.org/en/MediaCentre/PressBriefings/Pages/38-nmsc94polar.aspx#.VZEmLGw1-Hs (accessed 9 September 2015); further documentation on the Polar Code, as well as presentations from a related workshop, can be found on the IMO website, available at http://www.imo.org/en/MediaCentre/HotTopics/polar/Pages/default.aspx (accessed 9 September 2015).
6. For an overview, see UNCTAD (2009), pages 16–18; see also Economic Commission for Europe (2013), pages 15 and 41–43.
7. See the IMO press release, available at http://www.imo.org/MediaCentre/HotTopics/polar/Pages/default.aspx (accessed 9 September 2015).
8. See the interview with the Secretary-General of IMO, published on 25 February 2015, available at http://www.arctic-council.org/index.php/en/resources/news-and-press/news-archive/992-interview-with-secretary-general-of-the-international-maritime-organization-imo (accessed 9 September 2015).
9. For further information, including on the status of ratification, see the website of the United Nations Division on Ocean Affairs and the Law of the Sea, available at http://www.un.org/depts/los/index.htm (accessed 9 September 2015).
10. SOLAS entered into force on 25 May 1980 and, as of 30 June 2015, had 162 States Parties representing 98.6 per cent of world tonnage.
11. Other conventions dealing with shipping safety include: the Convention on the International Regulations for Preventing Collisions at Sea, 1972; the International Convention on Load Lines, 1966 (Load Lines Convention); the International Convention on Safe Containers, 1972; the International Convention on Standards of Training, Certification and Watchkeeping of Seafarers, 1978 (STCW); and the International Convention on Maritime Search and Rescue, 1979. Non-mandatory codes and guidelines include: the International Maritime Dangerous Goods Code, 2006 (SOLAS, chapter VII); the International Code for the Construction and Equipment of Ships Carrying Liquefied Gases in Bulk Code (International Gas Carrier Code 1993) (SOLAS chapter VII); and the 2008 Intact Stability Code.
12. MARPOL entered into force on 2 October 1983 and, as of 30 June 2015, had 153 States Parties representing 98.52 per cent of world tonnage. While all contracting States to MARPOL are bound by annexes I (prevention and control of pollution by oil) and II (noxious liquid substances), not all contracting States have ratified or acceded to the other annexes. For further information, see the IMO website.
13. Other instruments dealing with ship-source pollution, whose provisions are also applicable in the polar regions, include the 2004 International Convention for the Control and Management of Ships' Ballast Water and Sediments (Ballast Water Convention); the Nairobi International Convention on the Removal of Wrecks, 2007 (Wreck Removal, 2007); the Convention on the Prevention of Marine Pollution by Dumping of Wastes and Other Matter (London Convention, 1972) and its 1996 Protocol; the 1990 International Convention on Oil Pollution Preparedness, Response and Cooperation, and its Protocol on Hazardous and Noxious Substances (OPRC/HNS Protocol, 2000).
14. Nairobi International Convention on the Removal of Wrecks, 2007; see UNCTAD, 2014a, pages 78–79.
15. The MLC entered into force on 20 August 2013 and, as of 30 June 2015, had 66 States Parties. For an overview, see UNCTAD (2013), page 104.

16 MARPOL Special Areas are certain waters that require, for technical reasons relating to their oceanographical and ecological condition and to their sea traffic, the adoption of special mandatory methods for the prevention of sea pollution.

17 MARPOL annex I, regulation 15.

18 MARPOL annex II, regulation 13.

19 MARPOL annex V, regulation 5.

20 MARPOL annex I, regulation 43.

21 The Antarctic Treaty System regulates relations among States in the Antarctic. The main instrument is the Antarctic Treaty, which was signed on 1 December 1959 and entered into force on 23 June 1961. The original Parties to the Treaty were the 12 nations active in the Antarctic during the International Geophysical Year of 1957–1958. As of 30 June 2015, the present total number of Parties is 52. The Treaty is supplemented by recommendations adopted at consultative meetings, by the Protocol on Environmental Protection to the Antarctic Treaty (Madrid, 1991) and by two separate conventions dealing with wildlife resources, the Convention for the Conservation of Antarctic Seals (London, 1972) and the Convention for the Conservation of Antarctic Marine Living Resources (Canberra, 1980). The Convention on the Regulation of Antarctic Mineral Resource Activities (Wellington, 1988), negotiated between 1982 and 1988, will not enter into force.

22 See the Protocol on Environmental Protection to the Antarctic Treaty (1991), which entered into force on 14 January 1998, annex IV, articles 5 and 6.

23 The Arctic is a shallow sea sometimes covered by multi-year ice or single-year ice and surrounded by land masses. The Antarctic is an ice-covered continent which is surrounded by a deep ocean. The Arctic has been home to native peoples, who have made their living from the environment, for thousands of years. The Antarctic has no permanent population of people. The Arctic is currently less protected by international law than the Antarctic. For more information, see Det Norske Veritas (2011).

24 For a background on the importance of this issue, see http://globallast.imo.org/ (accessed 9 September 2015).

25 Held 13–17 October 2014 and 11–15 May 2015, respectively.

26 For further details, see the *Review of Maritime Transport 2013*. It should be noted that the issue of possible market-based measures was not discussed at the sixty-sixth, sixty-seventh and sixty-eighth sessions of MEPC.

27 For a summary of the content of the regulations, see UNCTAD (2012b), pages 97–98; for an overview of the discussions on the different types of measures, see UNCTAD (2011a), pages 114–116.

28 A copy of the study and further information on the methodology are available at http://www.imo.org/en/OurWork/Environment/PollutionPrevention/AirPollution/Pages/Greenhouse-Gas-Studies-2014.aspx (accessed 9 September 2015).

29 MEPC 68/5/1 (Marshall Islands).

30 These include identification of the primary fuel for the calculation of the attained EEDI for ships fitted with dual-fuel engines using LNG and liquid fuel oil.

31 These make the guidelines applicable to phase 1 (starting 1 January 2015) of the EEDI requirements.

32 These make the guidelines "applicable to level-1 minimum power lines assessment for bulk carriers and tankers, and agreed on a phase-in period of six months for the application of the amendments".

33 As required by regulation 21.6 of MARPOL annex VI, at the beginning of phase 1 and at the midpoint of phase 2, the Organization shall review the status of technological developments and, if shown to be necessary, amend the time periods, the EEDI reference line parameters for relevant ship types and the reduction rate; see IMO (2015a), page 28.

34 The intersessional Correspondence Group on Further Technical and Operational Measures for Enhancing Energy Efficiency, established at the sixty-seventh session of MEPC. The report is available as document MEPC 67/WP.13. For further information on the deliberations and documentation, see IMO (2015a) page 34.

35	MARPOL annex VI came into force on 19 May 2005 and, as of 30 June 2015, has 82 States Parties representing 95.23 per cent of world tonnage. Annex VI covers air pollution from ships, including NO_x and SO_x emissions and PM.
36	See http://onlinelibrary.wiley.com/doi/10.1002/jgrd.50171/pdf (accessed 24 September 2015).
37	For various opinions on the impact of emissions of black carbon on the Arctic and on global climate change, see: documents MEPC 68/3/5 and MEPC 68/3/5/Corr.1 (the Russian Federation), presenting data on black carbon emissions from shipping in ice conditions of the Arctic seas adjacent to the Russian Federation territory; the assessment by these documents of black carbon emissions from ships operating in the Arctic in ice conditions is that their impact is only regional and cannot pose a threat to climate change, and that black carbon emissions from ships can influence ice and snow properties only in cases where the emissions occur less than 100 kilometres from the ice edge; MEPC 68/3/19 (CSC), providing comments on document MEPC 68/3/5, pointing out that it does not follow any scientific standard for citations and assessment of differences to previous studies; and MEPC 68/3/22 (Norway), providing comments on document MEPC 68/3/5, requesting MEPC to continue its work on black carbon in accordance with the work plan agreed at MEPC 62.
38	For further discussion, see IMO (2014b), pages 35–39.
39	The first two SO_x ECAs, the Baltic Sea and the North Sea areas, were established in Europe and took effect in 2006 and 2007, respectively. The third established was the North American ECA, taking effect on 1 August 2012. In July 2011, a fourth ECA, the United States Caribbean Sea, was established. This latter area covers certain waters adjacent to the coasts of Puerto Rico (United States) and the United States Virgin Islands, and took effect on 1 January 2014.
40	Required under of MARPOL annex VI, regulation 14 "Sulphur oxides (SO_x) and particulate matter".
41	Consisting of 13 member States, one intergovernmental organization and six international non-governmental organizations.
42	For more information, see IMO (2015a), page 25.
43	Intended as a tool to assist in managing requests for spill response resources and offers for assistance from other countries and organizations when confronted with significant oil spill incidents.
44	Parts I (Basic information) and II (National policy) of the IMO Dispersant Guidelines have already been approved and will be published together with Part III. Part IV, covering subsea dispersant application, is under development and will take into account the experience gained from the Deepwater Horizon incident as well as other related technical developments.
45	See http://globallast.imo.org/ (accessed 9 September 2015).
46	Four proposed by the Republic of Korea and two by Singapore.
47	Two proposed by Japan and two by the Republic of Korea.
48	During 2014 and 2015, five States, Georgia, Japan, Jordan, Tonga and Turkey, acceded to the Convention.
49	For reasons related to the language and substance of this non-binding resolution, the delegation of the United States reserved its position with regard to it.
50	Initiated during the sixty-seventh session of MEPC and being conducted by the IMO secretariat in partnership with the World Maritime University.
51	With the collaboration of IMO and the International Oil Pollution Compensation Funds and International Tanker Owners Pollution Federation secretariats. Its purpose would be to promote the Convention by focusing on its fundamental public policy intent and objectives, rather than serving as a guide on how to implement the Convention.
52	For more information, see IMO (2015g), page 4.
53	Also highlighted in the *Review of Maritime Transport 2013*, pages 110–111; for further information on the international liability and compensation framework for ship-source oil pollution see also UNCTAD (2012a).
54	The 1992 Civil Liability Convention and 1992 International Oil Pollution Compensation Fund Convention; for an analytical overview of the legal framework, see UNCTAD (2012a).

55 See, for instance, "Shell Arctic oil drilling to commence within weeks", 3 July 2015, available at http://www.bbc.com/news/business-33379982 (accessed 9 September 2015).

56 For more information, see *On Scene Coordinator Report Deepwater Horizon Oil Spill*, submitted in September 2011, available at http://www.uscg.mil/foia/docs/dwh/fosc_dwh_report.pdf (accessed 24 September 2015).

57 See *Lloyd's List,* Is BP now an attractive takeover target? 2 July 2015.

58 Pillar I is mainly based on the model of the Container Security Initiative (CSI), introduced in the United States in 2002, and Pillar II is mainly based on the model of the C–TPAT, introduced in the United States in 2001. For more information on these, as well as for an analysis of the main features of customs supply chain security, namely advance cargo information, risk management, cargo scanning and authorized economic operators (AEOs), see WCO (2011). For a summary of the various United States security programmes adopted after September 11, see UNCTAD (2004).

59 As of June 2015, 168 out of 180 WCO member States have signed the letter of intent to implement the SAFE Framework.

60 A June 2012 version of the SAFE Framework can be found in WCO (2012). Also, the SAFE Package, bringing together all WCO instruments and guidelines that support its implementation, is available at http://www.wcoomd.org/en/topics/facilitation/instrument-and-tools/tools/safe_package.aspx_(accessed 24 September 2015).

61 The SAFE Framework AEO concept has its origin in the International Convention on the Simplification and Harmonization of Customs Procedures, as amended (revised Kyoto Convention), which contains standards on "authorized persons" and national programmes.

62 The first MRA was concluded between New Zealand and the United States in June 2007. As of June 2015, 32 bilateral MRAs had been concluded. A further 19 are being negotiated between: Brazil and the Republic of Korea; Canada and the European Union; Canada and Israel; Canada and Mexico; China and Israel; China and Japan; China and the United States; Costa Rica and Mexico; Costa Rica and the United States; the European Union and Hong Kong, China; Hong Kong, China and Japan; Hong Kong, China and Malaysia; Hong Kong, China and Thailand; India and the Republic of Korea; Israel and the Republic of Korea; Japan and Switzerland; New Zealand and Singapore; Norway and Switzerland; and the Republic of Korea and Thailand.

63 Due to the fact that 28 European Union countries have one common uniform AEO programme.

64 This is according to information provided by the WCO secretariat. For more information, see WCO (2015b).

65 These were Armenia, Azerbaijan, Colombia, Egypt, Georgia, India, Malaysia, Mongolia, Saudi Arabia, Serbia and Sudan. Furthermore, an AEO global conference was organized in Madrid in April 2014 and in spring 2016, a new global conference is planned for Mexico.

66 Regulation (EC) 648/2005 and its implementing provisions.

67 See, in particular, UNCTAD (2011a), which provides an overview of the major changes this amendment introduced to the Customs Code at pages 122–123.

68 For more information, see http://ec.europa.eu/ecip/security_amendment/index_en.htm (accessed 25 September 2015).

69 According to information provided by the European Commission's Taxation and Customs Union Directorate General, as of 11 June 2015, 17,782 applications for AEO certificates had been submitted and 15,476 certificates issued. The number of applications rejected up to 11 June 2015 was 1,881 (11 per cent of the applications received) and the number of certificates revoked was 1,383 (9 per cent of certificates issued). The breakdown reported per certificate type issued was: AEO-F 7,742 (50 per cent); AEO-C 7,152 (46 per cent); and AEO-S 582 (4 per cent).

70 For the self-assessment questionnaire, see http://ec.europa.eu/taxation_customs/resources/documents/customs/policy_issues/customs_security/aeo_self_assessment_en.pdf (accessed 25 September 2015). Explanatory notes are also available at http://ec.europa.eu/taxation_customs/resources/documents/customs/policy_issues/customs_security/aeo_self_assessment_explanatory_en.pdf (accessed 25 September 2015).

CHAPTER 5: LEGAL ISSUES AND REGULATORY DEVELOPMENTS

[71] The European Union has already concluded MRAs with Andorra, China, Japan, Norway, Switzerland and the United States. Negotiations are ongoing with Canada. For further information on AEOs, see http://ec.europa.eu/taxation_customs/customs/policy_issues/customs_security/aeo/index_en.htm (accessed 25 September 2015).

[72] See the European Commission press release of 21 August 2014 – Customs: Commission adopts strategy and action plan for better customs risk management, available at http://europa.eu/rapid/press-release_IP-14-936_en.htm (accessed 25 September 2015).

[73] See http://www.cbp.gov/border-security/ports-entry/cargo-security (accessed 25 September 2015).

[74] For more information on the various security initiatives, see UNCTAD (2004).

[75] For more information about CSI, see http://www.cbp.gov/border-security/ports-entry/cargo-security/csi/csi-brief (accessed 25 September 2015). The implementation of legislative requirements to scan 100 per cent of all United States-bound containers was again deferred in 2014 for another two years. See also UNCTAD (2014a), pages 86–87.

[76] The nine countries/territories are Canada, Israel, Japan, Jordan, Mexico, New Zealand, the Republic of Korea, Singapore and Taiwan Province of China.

[77] Importer Security Filing and Additional Carrier Requirements.

[78] For more information on the "10+2" rule, see http://www.cbp.gov/border-security/ports-entry/cargo-security/importer-security-filing-102 (accessed 25 September 2015).

[79] For more information, see http://www.cbp.gov/trade/isa/importer-self-assessment (accessed 9 September 2015). For information on the benefits for the participants, see http://www.gpo.gov/fdsys/pkg/FR-2002-06-17/pdf/02-15308.pdf (accessed 25 September 2015).

[80] For more information, see http://www.gpo.gov/fdsys/pkg/FR-2014-06-16/pdf/2014-13992.pdf (accessed 25 September 2015).

[81] For more information, see http://www.state.gov/t/isn/c10390.htm (accessed 25 September 2015).

[82] For more information, see http://www.cbp.gov/border-security/international-initiatives/international-agreements/cmaa (accessed 25 September 2015).

[83] On 1 July 2004, the 2002 amendments to SOLAS and the new ISPS Code entered into force and became mandatory for all SOLAS member States. For more information, see UNCTAD (2004 and 2007).

[84] See also *Lloyd's List*, 2015, Minimal changes made to ISO 28007 standards for maritime security, March, available at http://www.lloydslist.com/ll/sector/ship-operations/article459421.ece (accessed 9 September 2015).

[85] For more information, see IMO (2014g). See also http://www.imo.org/OurWork/Security/WestAfrica/Pages/WestAfrica.aspx (accessed 25 September 2015). Reports of actual and attempted attacks by pirates and armed robbers are promulgated via the Global Integrated Shipping Information System, available at http://gisis.imo.org (accessed 9 September 2015).

[86] Currently located in the Regional Maritime University in Accra.

[87] By the Economic Community of Central African States, the Gulf of Guinea Commission and member States in the region.

[88] Available at http://www.imo.org/OurWork/Security/WestAfrica/Documents/code_of_conduct%20signed%20from%20ECOWAS%20site.pdf (accessed 25 September 2015).

[89] For further information, see the Maritime Trade Information Sharing Centre website, www.mtisc-gog.org (accessed 9 September 2015). The newly updated *Guidelines for Owners, Operators and Masters for Protection against Piracy in the Gulf of Guinea Region*, developed jointly by the Baltic and International Maritime Council, the International Chamber of Shipping, INTERTANKO and INTERCARGO is also available via the IMO website as well as on the websites of those organizations.

[90] From China, Japan, Nigeria, Norway, the United Kingdom and most recently Angola.

[91] For IMO guidance on piracy and best management practices, see http://www.imo.org/OurWork/Security/PiracyArmedRobbery/Pages/Default.aspx (accessed 25 September 2015).

[92] The full text of the statement is set out in IMO (2014h), annex 29.

[93] The group, known as "the Kampala Process", comprises members of the Somali Contact Group on Counter-piracy and was established in 2010 with the objective of promoting coordination and information-sharing between counter-piracy offices of the Government of Somalia, Galmudug, Puntland and Somaliland.

[94] EUCAP Nestor is a European Union civilian mission, with some military expertise, under the Common Security and Defence Policy. EUCAP Nestor is an unarmed capacity-building mission with no executive powers, which aims to support the development of maritime security systems in the Horn of Africa and the western Indian Ocean States, thus enabling them to fight piracy and other maritime crime more effectively. For more information, see https://www.eucap-nestor.eu (accessed 9 September 2015).

[95] By the Special Adviser to the IMO Secretary-General on Maritime Security and Facilitation.

[96] For more information, see IMO (2015k), page 7.

[97] See, for example, ILO press release of 4 April 2014, ILO, "Maritime sector to address abandonment of seafarers and shipowners' liability", available at http://www.ilo.org/global/about-the-ilo/media-centre/press-releases/WCMS_240418/lang--en/index.htm (accessed 29 September 2015).

[98] Issues related to ILO Convention No. 185 on Seafarers' Identity Documents (Revised) 2003 were discussed during a Tripartite Meeting of Experts held on 4–6 February 2015.

[99] For more information, see http://www.ilo.org/global/standards/maritime-labour-convention/events/WCMS_301223/lang--en/index.htm (accessed 25 September 2015).

[100] Conducted by Seafarers' Rights International.

[101] The coefficient of determination, R2, between the HDI and the number of measures notified as category A is 0.3748, suggesting that about 37.48 per cent of the variation in the number of category A notifications per country is statistically explained by the country's HDI. Similar R2s are obtained for the correlation between the category A notifications and the GDP per capita (R2 = 0.36) and the share of individuals with access to Internet (R2 = 0.35). The number of trade facilitation measures notified as category A have been calculated by UNCTAD on the basis of the individual notifications published on the WTO website, available at http://www.wto.org/english/tratop_e/tradfa_e/tradfa_e.htm#notifications (accessed 25 September 2015). In several cases a WTO member notified specific measures as partially under category A; in these cases UNCTAD counted the case as 0.5. GDP per capita has been estimated by UNCTAD. Data are for 2013. The HDI is sourced from UNDP, available at http://hdr.undp.org/en/content/human-development-index-hdi (accessed 25 September 2015). Data are for 2013. The percentage of individuals using Internet is sourced from the International Telecommunication Union, available at http://www.itu.int/en/ITU-D/Statistics/Pages/stat/default.aspx (accessed 25 September 2015). Data are for 2013.